Dedicated to my Wife and Children

Introduction 6

Chapter 1 : The Pregnancy 10
Pregnancy and Drugs 11
Amniotic fluid 13
Umbilical cord 13

Chapter 2 : The Baby 16
At the delivery 16
The baby after birth 18
Examination of the newborn baby 20
Voiding and Stooling 31
Vitamin K 32
Haemorrhagic Disease of the Newborn 33

Chapter 3 : Early development, senses and reflex activity 34
Early development 34
The Senses 35
Reflex activity 37
Laterality 40

Chapter 4 : The first week 41
Nuchal Folds 41
Amnionic bands (Constriction bands: Streeter's bands) 41
Orbicularis oris hypoplasia 42
Facial nerve palsy 43
Brachial plexus injury: (Erb's and Klumpke's palsy) 43
The Skin 44
Sebaceous gland hyperplasia 45
Milia 45
Miliaria 46
Toxic erythema of the newborn: (flea bite dermatitis) 46
Transient Neonatal Pustular melanosis 47
Sucking blisters 47
Cutis aplasia 47
Ischaemic ulcer 48
Caput succadeneum 48
Cephalhaematoma 48

Chapter 5 : Birthmarks and Skin Rashes 50
Haemangiomas 50
Other Birth Marks 53
Skin rashes 55

Chapter 6 : The Head 61
The Fontanels 61
Skewed and flat heads 63

Chapter 7 : The Mouth 65
Cleft Lip and Palate 65
Sub mucous cleft 66
Uvula 66
Epstein Pearls 66
Congenital Epulis 67
Ranula (Salivary mucocoele) 67
Tongue tie (Ankyloglossia) 67
The Teeth 68

Chapter 8 : The Ears 72
Low set ears 72
Small and malformed ears 72
Vertical groove in ear lobe 73
Crumpled or cupped (lop) ears 73
Pre-auricular dimples (pits): 73
Pre-auricular skin tags 73

Chapter 9 : The Eyes 74
Brushfield spots 75
Wolfflin nodules 75
Cataracts 75
Coloboma 76
The Naso-lacrimal duct 76

Chapter 10 : The Chest 78
Breasts 78
The Heart 79
Cyanotic congenital heart disease 80

Chapter 11 : The Limbs 85
  Arms and hands 85
  The Legs and Feet 87
  Congenital dislocation of the hips (CDH) 89

Chapter 12 : The Genitalia 94
  Males 94
  Females 101

Chapter 13 : The Back, Abdomen and Anus 103
  The back 103
  The abdomen 104
  Inguinal hernia 105
  The Anus 106

Chapter 14 : Jaundice 108

Chapter 15 : The Breasts & breast feeding 118
  Breast feeding and drugs 123

Chapter 16 : Vomiting, posseting and spilling 125
  Gastro-oesophageal reflux 125
  Pyloric stenosis 128

Chapter 17 : Milk protein sensitivities 130
  Colic (infantile colic, evening colic, three months colic) 130
  Cow's milk protein allergy or intolerance 132
  Carbohydrates 135

Chapter 18 : Some symptoms and signs 137
  Bowel motions 137
  Breast tissue 138
  Breathlessness 138
  Cheeks 139
  Coughing 139
  Dehydration 139
  Diarrhoea 140
  Dummies (pacifiers, soothers, binkies) 140
  Ears 141
  Eyes 141
  Fontanelle 142
  Jaundice 142
  Length 142
  Lethargy 142

Rash 143
Sleeping 143
Smoking 144
Speech 144
Teething 144
Temperature 145
Urine 145
Vomiting 145
Weakness of a Limb 146
Weight gains 146

Chapter 19 : Conditions for discussion or investigation 147
Foetal alcohol syndrome 148
Down syndrome 149
Thyroid metabolism 151
Congenital hypothyroidism 151
Congenital hyperthyroidism 152
Haemophilia 153
Von Willebrand's disease 154
Thrombocytopenia 154
Meningitis 155

Chapter 20 : Possible early infections for baby 156
Group B Streptococcal infection 156
Staphylococcal skin lesions 157
Rubella (German measles) 158
Varicella (Chickenpox) 159
Neonatal Herpes 161

Chapter 21 : Behaviour 163
Biological instincts 163
Social instincts 164
Disciplines 165
Sleeping 167
Eating 169

Chapter 22 : Practical advances in the care of babies 171
Potable water and breast feeding 171
Vaccination 171
Resuscitation techniques 172
Jaundice 173

Index 174

Introduction

I was first introduced to the discipline of Paediatrics in 1966 and became intrigued by the wide range of acute and chronic problems that made up this relatively young and demanding discipline; medical, surgical, behavioural and social. Eventually in one form or another it became not only my occupation but also an abiding medical interest and this was especially so for those small trusting vulnerable non-blinking solemn babies who had recently been born. In the past apart from general Paediatric practice I have had a special interest in sub-specialties such as Syndrome identification and classification, Paediatric Neurology, Infant and Child behaviour and of course the Newborn baby. Babies are very satisfactory; they are easy to pick up and put down and they do not talk or answer back and they don't dissemble. In other words they are as you see them, trusting, innocent and unblemished. My practice was in both the Public and Private sectors. Private practice required me to be on call every day and night of the week and at any time; Frequently to attend deliveries to resuscitate the asphyxiated or potentially "flat" baby. This book apart from a chapter on behaviour in the sometimes difficult early years, encompasses the first few months of the newborn baby's life and while it is not all embracing it is an attempt to answer questions that may concern and perplex the new mother or possibly her midwife. It has been based on the experience of thousands of babies over the years as well as extensive reading. It is written mainly for mothers and midwives but could be helpful for medical students who have an interest in babies.

In my time there have been, not surprisingly, several important refinements in the management and care of the newborn baby and infant. The advantages of breast feeding have been re-affirmed and

Introduction

encouraged in a more vigorous manner and the assertions of Truby King in the early 1900's are still largely being followed. He advocated breast feeding for at least the first 9 months of life and the "humanizing" of cow's milk if it was to be given to babies. Nowadays proprietary formulae are made to resemble as closely as possible all constituents and their proportions as are found in breast milk. By so doing over several years Sir Truby King was responsible for the expected National infant death rate mainly due to summer diarrhoea (scouring) of one infant per day to fall to almost nil. There is little doubt that "breast is best" in terms of nutrition, health and intelligence.

The immunisation program is constantly being adjusted and expanded. In the 18th century smallpox was a scourge and a killer. The first attempts at vaccinating against this disease consisted of pricking the serum from the sores of smallpox sufferers into the skin of an un-infected person in the hope that resistance would ensue at the expense of only a mild case of the resulting illness. The father of all subsequent vaccination programs was Edward Jenner (1749-1823) who in 1798 treated smallpox by inoculating subjects (who must have shown admirable bravery and trust) with fluid obtained from the sores of vaccinia, a disease of cattle which resembled smallpox. This vaccination had none of the dangers of the earlier attempts with smallpox material and did confer immunity against smallpox. Louis Pasteur (1822-95) discovered the "germ" theory to explain diseases and then later went on to establish the fundamental principle that the injection of weakened (attenuated) cultures of a "germ" (organism) would give protection against the disease caused by that "germ". Thus the works of Jenner and Pasteur were the earliest examples of the production of passively acquired immunity. Subsequently there has been the discovery of more and more viruses, the development of antibiotics and an ever expanding immunological field.

Vaccination and the understanding of natural and passively acquired immunity along with antibiotics have proved to be two of the cornerstones in the modern medical management of babies and infants. These days the failure to use antibiotics for known susceptible illnesses or the failure, unwillingness or even negligence to have baby and child immunized would be seen as an abrogation of parental duties and care

The manner of resuscitation of the asphyxiated baby has also changed over the last two generations. In the past the simple and undignified expedient of holding baby up by the heels after the delivery was considered sufficient; later a few drops of Vandid on the tongue encouraged baby to gasp and intra-nasal oxygen might be given. More recently bagging with oxygen was introduced. For the "flat" limp, blue-white asphyxiated baby suctioning, intubation, oxygen and the administration of appropriate drugs are now standard management. From my point of view the most striking change was with the administration of intravenous Sodium bicarbonate. An apparently lifeless baby would almost immediately exhibit spontaneous respirations, become pink and develop a normal heart rate. Sodium bicarbonate reversed the metabolic acidosis which was the result of asphyxiation and hypoxia. This along with intubation and oxygen and if necessary intravenous glucose would usually produce an active pink(well perfused) baby.

Neonatal care and technology have made huge strides and many years ago when a very small baby would usually fail to survive, it is now not uncommon for the very smallest and earliest of gestational age babies to live and in many cases without the development of later deficits.

The treatment of jaundice has been simplified with benefits for both baby and mother. Prior to the introduction of Rhogam (anti D hyperimmune gamma globulin), exchange transfusions of baby's blood because of significant jaundice were common, time consuming, often alarming and also carried some morbidity. Prior to the introduction of Rhogam one hundred or more exchange transfusions in a year in Christchurch would be performed and often a severely jaundiced baby would require several sequential exchanges in the course of a day. Today an exchange transfusion for jaundice due to Rhesus incompatibility is uncommon. The advent of light treatment with the arrival of the overhead phototherapy unit and later the introduction of the fibre-optic blue light blanket simplified the management of jaundice. Previously these babies were managed expectantly while bilirubin levels steadily rose until it was time for an exchange transfusion and of course in that case transfer to another hospital or ward with the subsequent separation of baby from mother. The overhead phototherapy unit negated the need for transfer of babies since there was a phototherapy unit in all nurseries but it did require the bandaging of baby's eyes for protection. Fibre-optic blue light blankets now allow baby to be treated next to mother's bed and also allow the eyes to be left uncovered.

Thus more vigorous promotion and better understanding of breast feeding, better and more effective resuscitation techniques and the later neonatal management, the simplified rational treatment of jaundice, the use of antibiotics, and the benefits of immunization have been some of the most visible advances in care over the last two generations.

Chapter 1 : The Pregnancy

During her pregnancy a healthy woman should gain roughly about 10-12kgs (25lbs) in weight – one time when putting on weight is good. In the first two trimesters (3 months per trimester; ie 6 months) the weight increase that the expectant mother gains goes mainly towards the supporting infrastructures for the baby, namely maternal blood, extracellular fluid, tissue and fat stores and the placenta. During the last trimester the weight gain is directed towards the baby (foetus).

Foetal movements are seen as a reflection of the baby's health during the pregnancy. Foetal movements are due mainly to the strength of the lower limb movements and vigorous or sustained activity is the result of leg and trunk mobility. These movements are called "kicks", "stretches" or "roll overs." Hiccups are due to strong foetal diaphragmatic contractions. As early as the 16th week of the pregnancy mother may feel movements (quickening) but these are usually infrequent. On the other hand foetal movements should not decrease in frequency during the last week prior to delivery. If they do this suggests that there is foetal distress or some other antepartum complication. The foetus will have periods of activity and inactivity each day but periods of inactivity which last an hour or more do not necessarily reflect a physiological rest period. A loss of foetal movement despite the preservation of heart sounds may presage an intra-uterine death. This lack of movement which should be seen as an alarm signal can occur from twelve hours to four days before any foetal heart rate changes are detectable.

Two thousand years ago Hippocrates said that mothers carrying a female foetus had a pale face while mothers carrying a male foetus had a healthier tone to the skin. Excessive vomiting (which is often accompanied by facial pallor) during the pregnancy (hyperemesis gravidarum) is associated with raised concentrations of human chorionic gonadotrophin hormone (hCG) which is a hormone produced by the placenta. In a normal pregnancy with a female foetus, hCG levels are higher than in a pregnancy with a male foetus.

During the past two or so decades multiple gestation pregnancies have increased because fertility enhancing procedures and treatments have multiplied. In addition there has been a rise in the pregnancy rate in women over 35 years of age and older women are more likely to conceive twins.

Pregnancy and Drugs

A small percentage of congenital malformations seen in the baby are caused by drugs or medicines which have been taken during the pregnancy.

If a definite link can be established between the drug taken and the abnormality detected then the drug is called a teratogen. Certain conditions must prevail before a teratogenic effect is seen.

(a) exposure to the drug should occur at a critical time of embryogenesis. ie: during the first twelve weeks of the pregnancy.

(b) the offending agent must cause a specific set of malformations (embryopathy).

(c) teratogens are species specific and teratogenic effects cannot therefore be extrapolated from animals to humans.

(d) there is a genetically determined predisposition of either mother or baby which will determine the teratogenic potential of the drug or medicine.

Of all the agents that could cause problems if taken during the pregnancy many or most can be avoided; eg; drugs, alcohol, smoking and caffeine. The safest policy of all is if possible to avoid taking any drugs during pregnancy or to avoid pregnancy if taking

Chapter 1 : The Pregnancy

drugs. "nothing in life is to be feared. It is only to be understood" – Madame Curie.

There are some known teratogens which can cause congenital abnormalities:

(a) Alcohol - binge or heavy drinkers and alcoholics. - foetal alcohol syndrome.

(b) Warfarin – (anticoagulant) - bone dysplasias; deafness and eye defects

(c) Phenytoin, Carbamazepine, Tridione. - (anticonvulsants) - cranio-facial, 5th finger and nail abnormalities; cleft palate; tracheo-oesophageal fistula.

(d) Valproate - (anticonvulsant) - spina bifida

(e) Propylthiouracil - (anti-thyroid medication) - goitre.

(f) Cytotoxic agents - microcephaly; growth retardation; cleft palate.

(g) Lithium - heart defects.

(h) Thalidomide - limb and heart defects.

(i) Retinoic acid - Central nervous system defects; cleft palate.

(j) Steroid hormones - virilization.

(k) Streptomycin - hearing deficit.

(l) Tetracyclines - discoloured teeth.

There are some potential environmental hazards that may occur during the pregnancy. These hazards include smoking whether actively or passively and which can produce a growth retarded baby. Coffee (caffeine) intake should be limited to no more than three cups per day (there is about 100mgm of caffeine/cup) or otherwise replaced by de-caffeinated coffee since with excessive caffeine there is an increased risk of prematurity.

Certain processed cured meats (bacon) contain preservatives in the form of nitrosamines which can be responsible for brain tumours in the baby. Remember that all sedatives and analgesics taken in the pregnancy will appear in the foetus to some extent.

Foetal drug addiction is to be expected if heroin, morphine, methadone, cocaine, marijuana etc are taken during this time. A drug addicted baby will show withdrawal symptoms from birth and for the first hour or two these symptoms are manifested by a piercing high pitched cry, irritability, tremors, vomiting, sneezing and respiratory distress.

Amniotic fluid

Amniotic fluid is housed within the amniotic cavity. Amniotic fluid is the lake in which the baby bathes and provides a supportive medium where the foetus is able to move about freely. It also provides a protective buffer protecting the foetus from external injury. During pregnancy the amniotic fluid volume increases from about 30mls at ten weeks gestation to 1000mls at thirty seven weeks. After sixteen weeks of the pregnancy the major source of amniotic fluid is from foetal urine with some from the foetal lungs. Near term the foetus swallows 400-500mls of amniotic fluid daily (125mls/kg/day).

If there is less than 500mls of amniotic fluid present this is known as oligohydramnios while quantities greater than 2 litres are known as polyhydramnios. Amniotic fluid is a weak solution which contains urea, some inorganic salts, and a small amount of protein and glucose. If meconium is present in the amniotic fluid this means that at some time during the pregnancy and especially prior to delivery there has been foetal distress.

Babies are attracted to the odour of amniotic fluid and apparently prefer an amniotic fluid coated nipple.

Umbilical cord

The umbilical cord has an average length of 55cms with a wide range of variation from 30-100cms. The umbilical cord contains two umbilical arteries, a single umbilical vein, and the remnants of the connections with the bowel and bladder (vitelline duct and the urachus respectively) all of which are surrounded by and contained in a mucoid form of foetal connective tissue called Whartons jelly. There are no nerve fibres in the cord so that cutting the cord is painless for baby. A thick cord indicates that the baby has been well nourished from a healthy placenta, while a very thin cord suggests that baby has been nutritionally compromised. The cord is spiral shaped and this is thought to be because of unequal growth of the two umbilical arteries.

A single umbilical artery is found in about 1% of single births and in about 7% of twin births. Unless other abnormalities are present in the baby a single umbilical artery has no significance and does not call for any investigations.

The separation of the umbilical cord from the baby occurs later in babies that have been delivered by Caesarean section than in those babies delivered vaginally. This is probably because there is less bacterial contamination of the baby's skin after a Caesarean section when compared with a vaginal delivery and as a result fewer white cells are attracted to the cord. Cord separation is mediated by white cell (macrophage) infiltration with subsequent digestion of the cord. The time for separation of the cord can range from 3-45 days but any delay over 3 weeks raises the possibility of a defect in white cell function which causes a delay in macrophage aided digestion of the cord. Thus the practice of applying anti-bacterial agents to the cord will lead to a two-fold increase in the time interval before there is separation of the cord.

A delay in the separation of the cord (5-8 days after birth) or a cord infection indicated by a flame shaped area of reddening of the skin just above the umbilicus can later result in the proliferation of granulation tissue at the umbilicus. Granulation tissue appears as a raspberry like pedunculated projection and is known as an umbilical granuloma. An umbilical granuloma has to be differentiated from everted intestinal mucosa which will have a similar appearance in the cord. A granuloma is solid while everted intestinal mucosa is soft and will permit the admission of a fine probe. A silver nitrate caustic pencil may be applied if the site is red and oozing blood but if the discharge is not blood stained then do not apply a caustic or styptic pencil. Any leakage from the umbilical stump other than blood will be clear in colour suggesting urine or else contain meconium or faecal material. These findings will denote a residual connection with the bowel or bladder and must not be ignored. They will require investigations and surgical intervention.

Chapter 2 : The Baby

At the delivery

Immediately after the delivery which is usually a happy and sometimes emotional time for mother and father as well as the attending staff, the baby's condition is noted and an Apgar score is allotted to the baby. The Apgar score is recorded at 1 and 5 minutes of age and is a subjective assessment of the baby's condition at those two times and should be made by an experienced observer. It is measured by assessing the following five parameters.

1. heart rate (below or above 100 beats/minute)?

2. respiratory effort [from no respiratory effort through to irregular shallow respiratory attempts and best of all vigorous respirations with crying]

3. muscle tone (absent movement, weakly passive or active movements)

4. colour   (pale blue-white, pink with blue extremities or generally pink).

5. reflex irritability (absent through to facial grimace or active avoidance)

Scores of 0, 1 or 2 are allotted in each of these 5 areas. Thus a maximum score of 10 is possible at both 1 and 5 minutes after delivery.

In 1952 Virginia Apgar, an American Anaesthesiologist (Anaesthetist) developed a scoring system which was later named after her and which evaluated the condition of the baby after birth.

Chapter 2 : The Baby 17

This scoring system identified infants who were depressed and required resuscitative measures. The re-evaluation at 5 minutes was to assess the infant's response to the resuscitative measures applied. The original premise of the Apgar system was to assess the condition of the newborn baby after delivery and to predict survival in the neonatal period (1st 28 days of life). It was not meant to forecast the intelligence quotient (IQ) later in life.

The Apgar score is influenced by the baby's gestational age with the heart rate being the least affected by gestational age and also the most important score. Respiratory effort, muscle tone and reflex irritability improve with increased gestational age while colour is the least reliable index at any gestational age.

In practice the Apgar score is applied retrospectively as a record of the baby's condition in the first 1-5 minutes after delivery. The five parameters can however be used prospectively immediately after the delivery and will then give a good indication of how much resuscitation the baby is going to require. For instance baby may require any of the following; no intervention; a little suctioning (with or without oxygen); suctioning and positive pressure by face mask with oxygen (Ambu or Laerdal bag); intubation with suctioning and oxygen and in addition if baby is "flat" various appropriate drugs (naloxone, bicarbonate, adrenaline, calcium, glucose).

Remember that the face mask system (Ambu bag or Laerdal mask) is a relatively inefficient system for delivering oxygen to the lungs with tidal exchange volumes being less than 1/3 that of intubation. The face mask system will rarely produce adequate alveolar ventilation (oxygenation of the lungs) but is effective in part because it stimulates reflex inspiratory efforts by the baby.

A "flat" or asphyxiated baby is defined as one that requires more than one minute of positive pressure intubation before the onset of satisfactory and sustained respirations.

If thick meconium is present in the liquor the baby must be suctioned through the vocal cords as soon as possible after the delivery. Suctioning on the perineum is not helpful, wastes time and is unnecessary. The presence of meconium in vertex presentations often indicates foetal distress especially if there have been alterations in the foetal heart rate. Thick meconium will usually mean a depressed ("flat") baby at birth. The difficulty is to determine the difference between thick or thin meconium and no real agreement has yet been reached since there are no firm criteria. In other words it is a subjective impression only. Meconium is first seen in the fifth month of gestation. It derives its dark green-black colour from bile salts and is free from bacteria, has little fat (lipid) present and no protein. It has a high water content. If aspirated into the lungs meconium will cause a chemical inflammation (pneumonitis). Because of oxygen insufficiency (hypoxia) which causes foetal distress the foetal bowel becomes ischaemic (deprived of oxygenated blood). The ischaemic bowel then has a brief period of increased peristalsis and this with relaxation of the anal sphincter leads to the passage of meconium in the hypoxic foetus.

Later once the respiratory and circulatory rates are satisfactorily established a naso-gastric tube should be gently passed via each nostril in order to aspirate stomach contents and meconium and also to exclude obstruction of the nasal passages and the oesophagus (conditions known as choanal and oesophageal atresia respectively). The anus should also be inspected after the delivery to make sure that it is patent. This of course will avoid much embarrassment later should it eventuate that meconium has not been passed.

The baby after birth

A new born baby whose birth weight falls between the 10th and 90th percentiles for gestational age is said to have undergone normal foetal growth and to have an appropriate weight for the gestational age. For a full term baby this weight lies between 2.5kg (5lb 8oz) and 4.3kg (9lb 8oz).

The description low birth weight is given to any baby regardless of the gestational age who has a birth weight less than 2.5kg (5lb 8oz).

A premature baby is any baby born at or before 37 weeks gestation regardless of the birth weight.

Babies with a birth weight below the 10th percentile for their gestational age are called small for gestational age (SGA), while

Chapter 2 : The Baby

babies with a birth weight over the 90th percentile for gestational age are deemed to be large for gestational age (LGA).

A small for gestational age baby is considered to have undergone intra-uterine growth retardation and there are two main reasons for being small for gestational age. There will have been either a reduction in the growth potential of the baby or a failure of the growth support systems

A reduction in the growth potential of the baby leads to a symmetrically small infant with respect to weight, length and head circumference. There will have been interference with the early proliferation of cell numbers (hyperplasia) in the growth phase, There are two possibilities, either intra-uterine infection or a chromosomal abnormality.

On the other hand a failure in the growth support system will lead to a light infant but length and head circumference are spared and will be in the normal range. This failure in trans-placental nutrition can be due to several influences but maternal illness, maternal nutrition, drugs, cigarette smoking or living at a high altitude will play a part.

A father's birth weight will also correlate with the birth weight of his baby. However the main factor in foetal growth apart from the genetic endowment given by the parents is the growth support system supplied trans-placentally by mother. With caloric deprivation hair growth can be interfered with and the growth retarded baby will have sparse hair with poor texture. Conversely the large for gestational age baby with a healthy placenta will usually have a luxuriant growth of hair (for some reason often black). Babies usually double their birth weight by five months although this may be achieved a little earlier in male infants. The birth weight is tripled by about fifteen months.

The placenta serves to remove heat from the foetus and ordinarily the foetal temperature will be slightly higher than the maternal temperature. If there has been placental insufficiency there will be impairment of heat transfer from baby to mother and the small for gestational age baby can then be born febrile. There are several other reasons also for baby becoming febrile in the first few days of life; sepsis, birth asphyxia, phototherapy using an overhead light unit in the treatment of jaundice, drug withdrawal, a large cephalhaematoma, an overactive thyroid gland or dehydration.

But the most important problem for the Paediatrician when confronted with a febrile newborn baby is to determine if baby is well or not – to differentiate the baby with sepsis from the baby with what is called dehydration fever when baby is not receiving enough milk. The baby with dehydration fever will usually have shown a greater than expected weight loss but will look well. Babies can normally lose up to 10% of their body weight in the first week but not in the first few days. Only about 1% of healthy newborn babies will have an armpit (axillary) or rectal temperature greater than 37.8C during the first four days of life. In normal babies the rectal temperature is about 0.6C higher than the skin temperature on the anterior aspect of the leg. With fever due to bacterial disease the rectal temperature is much higher than on the leg. Thus if the rectal to anterior leg skin temperature difference is 1-2C or more then the fever is more likely to be disease related than due to other causes. A fever of over 39C and a low white cell count suggests bacterial disease.

Examination of the newborn baby

Before the examination there are several important questions for baby's mother to answer. These questions are necessary so that certain familial conditions can be ruled out, but which if present mean that later on, appropriate investigations for the baby should be undertaken.

- did antenatal scanning during the pregnancy reveal any foetal anomalies?
- has anyone in either family been born with dislocated hips? has anyone in either family had an early (40's or early 50's) hip replacement?
- was baby breech for any considerable time in the pregnancy or delivered as breech?
- has anyone in either family been born deaf or had early unexplained deafness?

Chapter 2 : The Baby 21

- has anyone in either family had vesico-ureteric reflux (refluxing of urine from bladder to kidneys)?
- have there been any bleeding disorders in either family?

The examination is carried out preferably in the first 24-48 hours after delivery. The baby is completely undressed and is examined when lying supine in the cot. The overall appearance and colour is noted ie; pink (normal), healthy tanned appearance (jaundice), pallor (anaemia possibly due to blood loss), dusky grey (cyanosis) suggesting a congenital heart lesion or pulmonary problem. Duskiness of the hands and feet (acrocyanosis) is normal.

Starting with the head and face the ears are viewed, first looking at the general shape of the ears and ear folds as well as for the presence of skin tags or dimples (pits) in front of the ear. The presence of a skin tag means that a formal hearing assessment will need to be arranged at about 8 weeks of age because there can be a link between deafness and pre-auricular skin skin tags. The presence of unilateral or bilateral pre-auricular dimples is of no significance and is usually a familial trait. An ear may be grossly deformed or vestigial in which case baby will be deaf on that side. Usually there will be some cupping of one or both ears due to intra-uterine compression and this is a positional change which will resolve spontaneously. The speed of resolution can be gauged by the amount of cartilage in the upper part of the ear (helix) and tested by bending the ear forwards and then suddenly releasing. There will be a subjective impression of a rapid return to normal if there is normal elasticity and a slow return if there is reduced cartilage.

There may be facial asymmetry because the head and neck have been pushed to one side (torticollis). Again this is a positional change due to intra-uterine compression. If there is obvious facial asymmetry there will be cupping of the ear on the same side and a discernible indentation will be seen in the neck below the ear where the shoulder lay. A receding jaw (retrognathia) is indicative of intra-uterine compression because baby's neck has been flexed with the chin pressed against the chest. In this case there will be an inverted crease on baby's chin. Positional retrognathia may lead to some initial latching difficulties for the breast feeding mother. A receding chin can also be due to a familial trait in which case the term micrognathia is used.

The eyelids are usually swollen (oedematous) because of squeezing pressures from the delivery (no matter what sort of delivery). The tissues are loose in the eyelids and fluid extravasates easily into the lids. While the eyelids remain swollen the nasolacrimal duct which opens as a punctum in the corner of each eyelid and which drains the tears away will be blocked and the eyes will therefore appear wet or sticky. If the eyelids look sticky or messy, so long as the white part of the eye (sclera) remains white the eye itself will not be infected and simple eye toilet using expressed breast milk, saline solution or tap water is all that will be needed. If there is still concern or uncertainty then an eye swab is taken for culture.

The face may be congested and dusky with scattered pin point blood spots (petechiae) visible over the forehead and around the orbits. In association with this facial congestion the sclerae may show a blot or arc shaped haemorrhage on either side of the iris. Facial congestion happens after a rapid or difficult delayed delivery, or if the cord has been wound tightly round baby's neck. The blood vessels leak because they have been either distended or the pressure has been suddenly reduced.

When some babies cry the mouth will seem to move unevenly. This is because there is relative hypoplasia [limited growth] of one of the lower quadrant muscles around the mouth and lips (orbicularis oris group). Orbicularis oris hypoplasia can be confused with a facial nerve palsy when one side of the face and forehead have obviously limited movement and with an inability to completely close the eye on the affected side. With orbicularis oris hypoplasia the facial nerve is intact and the cheeks, forehead and eyelids move easily.

The nose may be deviated due to intra-uterine compression but this will correct spontaneously. The nose may have a yellow speckled appearance because of sebaceous gland hyperplasia – there are abundant fat glands in the nose.

Discrete white spots either singly or in groups will often be seen on the face and if present in crops will usually be noticeable on each

Chapter 2 : The Baby

side of the chin. They are called milk spots (milia) and they are small retention cysts which disappear after a few weeks.

Over the forehead sweat retention cysts (miliaria) appearing as clear blebs may become obvious after 24 hours and they will rupture leaving a peeling border. Miliaria are temporary and far less common these days because there is now less nursery humidity than in the past, humidity being a causative factor.

Pink flat birth marks with an irregular margin on the forehead (angel's kiss), above the nose (glabellum), and the eyelids are seen in many babies but are less common on the nose and upper lip. These birth marks all resolve and may take a year or more to do so depending on their size. Birth marks that can be easily blanched by gentle pressure will eventually disappear. Rarely a port wine stain (smooth edge, flat, uniformly red and unable to be blanched) will be present and port wine stains are permanent, unless subjected to laser therapy. Also seen although uncommon on the face are café au lait patches which have a pale tan appearance and which are also permanent.

The next step is to examine the neck looking on each side for the presence of dimples or clefts and which if present indicate a branchial cleft which will require further investigations such as ultrasound to determine the length of the associated track (fistula). These clefts if they leak will require surgical excision. In the midline the neck is palpated to exclude any suggestion of thyroid enlargement.

The hands are then examined looking for thumb or finger anomalies such as a bifid thumb, an extra digit which is usually an accessory little finger, soft tissue syndactyly (fusion) of the fingers or finger amputation and deformation caused by amnionic bands. An accessory little finger may hang by a thin pedicle in which case it can be ligated but if there is any suggestion of an extra bony ray or the pedicle is thick then formal excision when baby is older is undertaken. The 5th finger may be short and incurved (clinodactyly) and this is usually a family trait. Sometimes the 5th finger nail is smaller (hypoplastic) than expected and this feature can be associated with certain anticonvulsants taken during pregnancy

If the finger nails are markedly convex, more so than the normal convexity expected then look for other stigmata such as swollen oedematous feet and hands, shortened 4th fingers, webbing of the neck and widely spaced nipples in a small baby. The combination of these features are suggestive of Turner syndrome (sterile females).

The palms are examined with respect to the palmar creases which are commonly known as the heart line, head line and life line. A long proximal crease (head line) may extend right across the palm (Sydney line) and is familial and of no significance. A single transverse crease (Simian crease), the result of fusion of the distal and proximal creases (heart and head line) may be seen and again is usually a familial trait although it can be linked with a chromosomal abnormality such as Down syndrome. In that case there will be additional associated Down syndrome stigmata present. The hands are normally cool and dusky for the first few days since the circulation around the extremities of the baby is sluggish.

After the hands have been examined baby is lifted to a sitting position by holding the hands and pulling up with gentle traction. Muscle tone in the neck and upper limbs can then be assessed. Usually there will be some flexion of the elbows and only a slight degree of head lag in a baby with normal muscle tone. Once the baby is sitting and supported the skull is checked. Overriding of the skull bones at the various sutures will usually be felt and in the occipital area some sharp corners due to the shape of the occipital bone will be noted. Fusion of the sagittal suture giving a boat shaped skull or fusion of the coronal sutures giving a flattened skull posteriorly needs to be excluded. In these cases the sutures will be felt as a sharp ridge rather than a step due to overriding bones. An asymmetrical skull shape (plagiocephaly) is as a rule the legacy of intra-uterine compression and will resolve spontaneously. The anterior fontanelle at the junction of the parietal and frontal bones is palpated and should be soft and slightly depressed when baby is in the upright position. The posterior fontanelle at the junction of the parietal and occipital bones is small and may not be palpable. The so called third fontanelle is not a true fontanelle and is a notch in the sagittal suture behind the anterior fontanelle.

While baby is supported in the upright position the startle (Moro) reflex is elicited. A sudden alteration in the baby's position

Chapter 2 : The Baby

backwards from the vertical will cause the arms to be flung out briskly and symmetrically and the hands opened. If there is an asymmetrical response with one arm hanging limply or moving sluggishly and not as freely as the opposite arm, this will be due to a brachial plexus injury (Erb's paresis). With an Erb's paresis the hands will be able to close but if the injury is severe then the hand of the affected arm will not close (Klumpke's palsy).

The vast majority of Erb's pareses recover without any intervention, but in a more severe case after an interval of a week gentle passive stretching movements will help. If there is bilateral paucity of movement then consider bilateral Erb's pareses or a floppy (hypotonic) baby with generally poor muscle tone. With any brachial plexus injury the clavicles (collar bones) should be carefully palpated to exclude any suggestion of a grating sensation (crepitus) and which if present would indicate an associated fracture of the clavicle. Fractures of the clavicle recover without any intervention but baby should be nursed on the side away from the fracture to lessen pain.

After a week or two a bony lump (callus) will be felt over the site of the fracture indicating bony union with the formation of new bone and the disappearance of any discomfort. Once bony union has been achieved the lump will slowly disappear as remodeling of the clavicle takes place.

After the startle or Moro reflex has been performed baby will be upset and crying. At this stage however one needs a quiet baby and with the eyes open. Placing baby in the supine position (baby lying on the back) I insert the left index finger into the baby's mouth with the pulp of the finger (not the nail) against the baby's hard palate. Baby will automatically suck on the finger and quickly become quiet. This sucking reflex will also make baby open the eyes.

While baby is quiet and with the eyes open the heart, lungs and eyes can then be examined. The heart is checked for heart murmurs (not all are present at birth), the heart rate and for any variation in rhythm. If a heart murmur is audible, the peripheral pulses and the pulses in the groin (femoral pulses) must be found and felt. Also the area between the shoulder blades should be auscultated since the murmur of a coarctation of the aorta as suggested by absent or reduced femoral pulses will be noted there. Oxygen saturation levels in the blood should be checked if any heart murmur is detected.

On auscultation breath sounds should be easily audible and even over each side of the chest and with the finger still in baby's mouth the ease of baby breathing through the nostrils is observed. If there is any difficulty a fine catheter is passed through the nostrils to exclude narrowing or blockage of the nasal passages (choanal stenosis or atresia).

With the index finger still in baby's mouth the eyes will remain open and they are then examined using an ophthalmoscope. The shape of the pupils should be round and a tear drop shape would indicate a defect (coloboma) of the iris. An irregular black appearance at the periphery of the pupil where the iris has become everted may be seen – ectropion iris. This is common and normal. The lens is examined for the red reflex and if present then lenticular cataracts are unlikely. The retina is glimpsed but usually only a limited inspection is possible. Most Caucasian babies commence life with gray-blue coloured eyes and usually by 6-9 months the final colour will have become apparent. The pigment producing cells in the iris take a few weeks to develop and it is not uncommon for the eyes to later darken in colour.

With a finger still in baby's mouth the abdomen is palpated and should be soft. At the upper part of the abdomen and below but attached to the breast bone (sternum) a prominent lump may be seen and felt. This is the lower end of the sternum (xiphisternum) which is cartilaginous and bent upwards because of intra-uterine compression when the foetus was folded in the womb. The lower border of the xiphisternum is smooth but occasionally a notch may be felt in the border and this is a familial finding.

The abdomen is prominent in babies because the liver is relatively large and this coupled with a small pelvis causes the intra-abdominal organs to reside in the abdominal cavity. A very flat (scaphoid) abdomen should raise the possibility of a diaphragmatic hernia when some of the intra-abdominal contents will have been displaced into the chest (thoracic) cavity. A diaphragmatic hernia requires urgent surgery. The abdomen is palpated to determine if the spleen can be felt (usually it is not) but if it is then a haemolytic condition causing

Chapter 2 : The Baby

rapid destruction of red blood cells is the likely reason. The soft lower edge of the liver is normally palpable a few finger breadths below the costal (rib) margin.

If the bladder can be felt in a male infant then obstruction to urinary outflow because of urethral valves is probable. Umbilical stumps are fairly common and suggest that an umbilical hernia may develop later in which case it will be possible to feel a ring at the base of the umbilicus. Above the umbilicus there may be a long cylindrical protrusion in the abdominal wall where there is divarication of the recti abdomini muscles which have not yet fully united. This is normal, not a hernia and slowly resolves.

Lymph nodes may be palpable in a third of healthy newborns. They are usually in the inguinal area and in the newborn are no larger than 3mm in diameter. While feeling for lymph nodes the femoral pulses must be detected in the inguinal region – if they cannot be palpated usually the index fingers are in the wrong place but if not then congenital dislocation of the hips or coarctation of the aorta must be ruled out.

Very rarely an inguinal hernia may be present at birth. The anus is inspected. It should be centrally placed in the pigmented peri-anal area and not placed too far forwards. An imperforate anus is obvious and should be found at the time of the delivery. An anteriorly placed anus will probably mean some later difficulties with constipation. In females a perineal membrane may be present stretching from the fourchette to the anus. This membrane will slowly develop a skin covering (epithelialise) and will cause no problems. Occasionally a haemorrhoid is seen and this will slowly regress.

The genitalia are then examined. In males one is looking for hypospadias where the external meatus is not placed at the tip of the penis but may be seen further down the shaft. If that is the case the site from where urine is passed should be noted. There will be a natural partial circumcision in these cases but a natural partial circumcision can also be seen with no evidence of hypospadias. If a hypospadias is present usually the fibrous band (chordee) on the under (ventral) surface of the penis will be short and tight. The testes must be checked to ensure that they are palpable and are either fully or partially descended. If the testes are impalpable in a full term baby then an ultrasound study of the abdomen is required to locate them in the abdomen. If they cannot be seen then chromosomal studies should be undertaken. Sometimes a small white cystic swelling will be seen on the foreskin (preputial cyst) and this will rupture spontaneously in the first few days.

In females the labia are noted. In a full term or post mature baby the labia majora will be prominent and juxtaposed. In a pre-term baby the labia majora will be less prominent and more widely separated. Sometimes the labia minora are fused by a transparent membrane and this should later dissolve spontaneously but if it does not it can be divided surgically. A cystic swelling between the labia minora will probably be a hydrocolpos where secretions and debris accumulate behind the hymen. If the uterus is also involved the bulging cystic swelling is called a hydrometrocolpos. This swelling may rupture spontaneously or it can be incised surgically later. The clitoris should not be enlarged and the labia majora should not be fused. Fused labia with palpable gonads mean an intersex problem while fused labia with an enlarged clitoris but no gonads palpable will represent the adrenogenital syndrome

At this stage the crossed extension reflex is examined. With baby supine one leg is held down in an extended position and the sole of that foot is stroked. The position which the other leg takes up in response to the stroking stimulus will give an idea of the baby's gestational age – knee remains flexed (37 weeks or less); leg is extended (38 weeks); or leg is extended and crossed onto (39 weeks) or over the shin (full term or more).

The feet and ankles are then checked. Positional talipes or calcaneo-valgus deformations are common and reflect intra-uterine compression and they will resolve spontaneously. They can be easily passively over corrected and there is no suggestion of calf muscle wasting or asymmetry. With true (structural) talipes equino varus the foot and ankle cannot be over corrected passively and there will be evidence of calf muscle asymmetry with wasting. Calcaneo-valgus deformations are always positional and never structural. Soft tissue fusion (syndactyly) of the 2nd and 3rd toes is a not uncommon familial congenital anomaly without any significance and which passes from one generation to the next. Baby is now picked up and held vertically and facing away from the examiner. The calf muscles are checked for symmetry. Thigh creases are noted for symmetry but if asymmetrical have little clinical relevance. If there is asymmetry of the gluteal (buttock) creases this may be more important with respect to congenital dislocation of the hips. It used to be thought that asymmetry of the thigh creases meant congenital dislocation of the hips but this is a most unreliable sign and should not mean that ultrasound studies of the hips be arranged on the basis of this finding alone.

Sacral or coccygeal dimples are looked for. Coccygeal dimples are always normal and so are shallow sacral dimples. Sacral dimples are seen at the upper end of the sacral or gluteal cleft while coccygeal dimples are lower down in the cleft. If the base of the sacral dimple cannot be seen or if there is a surrounding birth mark or a protruding tuft of hair then a tethered cord (spinal dysraphism) must be excluded with appropriate investigations. The entire spine is checked to exclude any abnormal curvature. Usually a birth mark (stork bite) will be seen in the neck (nuchal) area and most of these birth marks can be blanched. About 10% of "stork bites" persist as Unna's naevus while the remainder slowly disappear over several years.

With the baby held vertically the placement or stepping reflex is performed by bringing the upper surface (dorsum) of baby's foot into contact with the underside of a table ledge. Baby will then lift and place that foot on the table and the manner in which the foot is placed is noted; toes first (37 weeks or less); flat of foot (38 weeks); front of foot (39 weeks) or back of heel (full term). This reflex coupled with the crossed extension reflex will give a reliable estimation of baby's gestational age down to the half week. Most babies place the right foot first but a small percentage persist in placing the left foot first. These babies will probably eventually become left handed and often there is a family history of left handedness in these instances.

Baby is then returned to the cot and the hips are examined. With baby in the supine position the pelvis is immobilized with one hand and the hips are checked. The thigh is flexed at right angles on the hip and firm downward pressure is exerted on the flexed knee. A subjective impression of thigh length and subsequent height can also be obtained but this depends on many babies and examinations

There are several possible findings when the hips are examined for dislocation – (a stable hip; an unstable (subluxable) hip; a dislocatable hip; or a frankly dislocated hip and a "clicky" hip when ligaments run over bony prominences) A "clicky" hip is of no significance. Much experience is needed before one can be certain about the stability of the hips. Dislocated hips are far more common in females but when present in a male there can be difficulty in obtaining a normal long term result even with treatment.

After the hip examination (which is uncomfortable if not painful) is completed baby is usually upset and crying. The opportunity is then taken to check the tongue and palate. The uvula at the back of the palate can be seen to be either intact or bifid. A cleft of the soft and/or hard palate is looked for and the palate is palpated to exclude a sub mucous cleft. If there is a sub mucous cleft a wide midline notch will be felt at the junction of the soft and hard palates. The palate must always be palpated if a bifid uvula is found. All babies will have a few pearly white cystic lesions (Epstein's pearls) in the mid-line at the junction of the soft and hard palate and these will disappear over the next few months. The tongue is checked and the unlikely event of true tongue tie excluded. The membranous fold (frenulum) beneath the tongue is normally visible and may reach the tip of the tongue. As the tongue grows longer and thinner the frenulum seems to recede. Contrary to popular opinion breast feeding is not interfered with in any way and if there are difficulties with breast feeding they should be looked for in other areas. In any event baby sucks with the body of the tongue and not the tip.

Sublingual retention cysts on the mouth floor (ranula) and congenital epulis (a swelling from the upper or lower jaw) are uncommon and are usually obvious.

After the examination of the hips and also after using a wooden tongue depressor and torch to examine the mouth and palate baby is usually distressed and crying. The final part of the examination is now carried out. Baby is lifted up and held in the air in a sitting position with the examiner's right hand beneath the buttocks and left hand clasping baby around the chest with baby's arms free. This position (the oculo-vertical reflex) will make baby open the eyes and fix and will also, depending on baby's personality and nature make baby stop crying. If baby has a settled temperament this will occur within a few seconds and indicates an equable nature. Babies who have a fractious unsettled temperament will be difficult to settle and will take several minutes to do so. This reflex gives one a good indication of baby's basic temperament and nature. Parents are reminded however that while the baby's nature is as demonstrated, nurture is equally important and will reflect management in the next few years. Most babies however start life with an even temperament and nature. This of course can be altered temporarily in the early weeks if baby develops colic or the pain of heartburn due to silent gastro-oesophageal reflux

At the end of the examination one should be able to offer a reliable opinion regarding baby's health, gestational age, the findings on the physical examination, the temperament and state of alertness and also some idea of the future with respect to later height and laterality

Voiding and Stooling

Nearly all (95%) babies will have passed meconium within the first 24 hours after birth and all babies should have done so within 48 hours. The passage of meconium may be delayed normally in premature babies. If there is delay in passing meconium look for an imperforate anus or an anterior anus. Abdominal distension with visible loops of bowel in association with bile stained vomiting indicates a probable bowel obstruction.

Meconium is the thick black-green material that is present in the lower small bowel (ileum) and the large bowel (colon). It consists of intestinal secretions, bile, desquamated cellular debris and amniotic fluid. A firm translucent conical plug is attached to and precedes the first meconium motion that is passed.

Most babies have passed urine within 24 hours after birth with 93% of babies voiding within 24 hours and 99% by 48 hours. At the time of delivery 25% of males will pass urine compared with 7% of females. The most likely reason for delay later in voiding is due to inadequate perfusion of the kidneys and spontaneous voiding should occur within 24-36 hours after commencement of feeding. A full bladder and failure to pass urine is usually due to maternal medication but in male babies with a poor stream urethral valves must be considered. The normal newborn baby will void 2-6 times during the first 48 hours and after that somewhere between 5-20 times a day. The urine is pale in colour but in the first few days may contain urates which turn a rose pink colour in air and leave a pink stain in the napkin. There is little ability to concentrate urine in the newborn but by four weeks the kidneys will be commencing to weakly concentrate urine.

Vitamin K

Vitamin K is needed in the first six months of life to prevent the rare but potentially catastrophic bleeding disorder, haemorrhagic disease of the newborn. Vitamin K comes in three forms

1. Vitamin K1 which is synthesised by leafy green plants;
2. Vitamin K2 a component of dairy products such as cheese and certain gut bacteria.
3. Vitamin K3 a synthetic water soluble form.

A Vitamin K deficiency exists in all newborn babies for several reasons. There is poor trans placental passage of Vitamin K from mother to baby with a ratio of 30:1 of maternal blood to cord blood; a sterile foetal gut which means the absence of Vitamin K producing bacteria; and an immature liver which would ordinarily concentrate Vitamin K that has been absorbed from the gut.

The recommended daily Vitamin K intake for an infant is 1mcg/kg/day. Breast milk contains about 2mcg/litre with twice as much in hind milk as fore milk. Proprietary milk formulae contain 30-60mcg/litre.

Haemorrhagic Disease of the Newborn

It is manifested in three forms:

1. Early – in the first 2 days of life.
2. Classical – from day 3 to day 7.
3. Late – from 8 days to 12 weeks.

In the 1950's before prophylactic Vitamin K was given, there was an incidence of 4/1000 fully breast fed babies developing haemorrhagic disease of the newborn, with most being either early due to certain maternal drugs or classical disease. In early or classical haemorrhagic disease there can be haemorrhages in the skin or gut, an enlarging cephalhaematoma or more serious and life threatening, haemorrhages into the liver, the adrenal glands or brain.

These can be prevented by a single dose of Vitamin K given to the baby at birth either by intramuscular injection or by mouth. The main problem now is to prevent the onset of late haemorrhagic disease which is rare and confined to breast fed babies. A single intramuscular injection of Vitamin K prevents all forms of haemorrhagic disease of the newborn but a single oral dose does not and therefore repeated oral doses are required. The current recommendations are that all newborn babies are given 1mgm of intramuscular Vitamin K at birth, but if this is refused by parents then 2mgm of Vitamin K should be given to the baby by mouth at birth, at 5 days and later at 6 weeks. Note that a single oral dose of Vitamin K lasts about 4 weeks and since late onset haemorrhagic disease can occur up to 3 months and sometimes 6 months of age then a case can be made out for further oral doses after 6 weeks. The minimal dose of Vitamin K that prevents bleeding is 0.025mgm and it can be seen that much higher doses are being given either orally or intramuscularly.

These days whether to administer Vitamin K or not to baby has become an emotional issue with common sense being the sacrificial victim. Every newborn baby should receive Vitamin K by intramuscular injection or by mouth shortly after delivery.

Chapter 3 : Early development, senses and reflex activity

Early development

Immediately after birth apart from visual fixing which is an indication of cortical neuronal function, all other activities which the baby displays are the result of reflex activity. The newborn baby can hear, smell, see, taste and feel from the outset although most movements are random and seemingly without purpose. Human responsiveness to sound begins for baby in the third trimester and babies are able to discriminate their mother's voices and may increase their sucking during times when they can hear their mother's voice. The normal newborn baby will turn the head to sound and will follow mother's face and move the body in rhythm to her spoken word. But apparently the newborn baby shows no preference for father's voice over any other male voice. Thus in the first few days after birth baby will prefer mother's voice over that of other women but will not demonstrate a similar preference for father's voice.

From an early age baby will recognize mother by smell (by end of first week), sound and later sight. The baby will stare fixedly at mother while feeding (one of the early bonds with mother) and will clutch tightly. Mother is alert to signals emitted by the baby one of the most obvious being the baby's cry and its quality. The cry may indicate hunger, feeling too cold or hot, tiredness or alarm due to separation. If baby is distressed there will be rapid aimless movements of the limbs, head and body and the skin may become flushed. On the other hand when baby is comfortable there will be quiet head turning, grasping of the hands, staring and soft cooing noises which will precede drowsiness and sleep.

Skin to skin contact within the first few minutes to hours after birth is said to facilitate maternal-infant bonding and while this is desirable and pleasurable many studies have suggested that there are no long term beneficial or detrimental effects for mother or infant. It is wrong to suggest that the mother who has not had early skin to skin contact in this way cannot be as competent or caring as the mother who did have early skin to skin contact.

Babies respond to the manner in which mother handles them. Most mothers whether primiparous or multiparous initially hold their baby in their left arm to the left of the midline regardless of whether they are right or left handed. This preference is also seen in non-pregnant women, new fathers and fathers with other children. Men without children exhibit much less left sided holding than do parents. The preference for left sided holding seems to develop during childhood in females and could be a genetically determined female behaviour. It is not seen in young boys and males without children. Right sided holding mothers show less body contact and can show a delay in accepting the newborn baby. In addition to left or right sided holding there is a third and more concerning possibility when baby is carried in the hands either facing away from or towards mother.

The Senses

Smell

Female babies are attracted by the smell of mother's areolar gland secretions, but male babies show no preference for mother's nipples over other mothers. Babies are also attracted by the smell of amniotic fluid and will prefer an amniotic fluid coated nipple. They imprint very early with respect to smell and this may reflect pre-natal exposure or pre-natal olfactory learning. The newborn baby can identify mother's odour by the end of the first week.

Taste

Newborn babies not surprisingly prefer sweet to non-sweet solutions and indicate their dislike for non-sweet solutions by facial grimacing and a change in expression. The basic non sweet flavours are salty, sour and bitter and their order of preference in descending order is salty, sour and lastly bitter. Sugar solution gives an initial transient negative reaction followed by facial relaxation and sucking. Sour tastes will cause lip pursing while bitter tastes will show mouth gaping. There is no consistent response to salt. The newborn baby can therefore distinguish between sweet and non-sweet tastes as well as between sour and bitter.

Infants receiving 2mls of a dilute (about 15%) sugar solution on the tongue before experiencing a painful procedure have a reduced crying time after three minutes compared with infants receiving a placebo. Sucrose is thus a useful and safe analgesic for minor painful procedures in babies. It appears to have some sort of opiate effect by releasing endorphins from the brain since the beneficial effects of sucrose can be ablated by anti-opioids (Naloxone).

Vision

Newborn babies do see and can respond to certain visual patterns and shapes. If offered a card with the choices of a human face, a scrambled human face or a blank card, the baby will turn more often to the human face suggesting that newborn babies are programmed for this recognition, rather than learning to recognize. Vision is a cortical response and suggests integrity of the central nervous system. True blinking in response to an object in the visual field develops between 2 – 5 months. The vision of the newborn is about a 20th – 30th of normal adult vision and adult visual acuity is reached by about 2 years of age. The newborn sees best those objects that are close and which have high contrast. The retina consists of the central fovea which is composed of cones and which are responsible for detail and colour while at the periphery of the retina are the rods which mediate movement and brightness. There is some uncertainty about colour recognition but by about eight weeks babies have red and green cones but blue cones apparently do not appear until about three months of age. Thus in the first month or two baby may have some insensitivity to blue colours. Babies establish normal ocular alignment by about eight weeks of age. Before that age alignment may shift from internal squinting (esotropia) to normal alignment or divergent squinting (exotropia).

For the first six weeks infants track with jerky eye movements but by three months smooth tracking is evident.

The iris continues to develop in colour for the first six months and the colour at birth which is usually blue-gray for Caucasian babies may become darker as the melanin pigment producing cells in the iris which develop later begin to produce pigment. The final colour is apparent in most babies by 6-9 months of age. Tears are present at birth but are not produced in response to crying for 1 – 3 months.

Reflex activity

Sucking (cardinal points) reflex

This is the most important reflex for the baby since it is essential for feeding and ultimately survival. It is also known as the cardinal points or rooting reflex and is seen when the lips or mouth are stimulated by touching or stroking. When the baby is sucking or breast feeding the eyes will usually be open presumably to see mother and thus helping to form a close bond between mother and baby. The vigour of baby can be assessed by the forcefulness of the sucking reflex and a baby with a well developed sucking reflex may also have sucked the hand or wrist during pregnancy and leave a series of blisters in varying stages of maturation. There is a strong need for sucking in the first 6 months of life, but after this time if a dummy has been used for soothing purposes, it becomes a habit although it may give some feeling of security. The use of a dummy is less harmful than thumb sucking but nevertheless should be discarded by 10 months at least. Most children have stopped thumb sucking by about 4 years. Between 4 and 14 years is the period of greatest dento-facial development and thumb sucking during this time may mean later orthodontic treatment and maldevelopment of the maxillae. Thus there is a stage of normal and clinically insignificant thumb sucking from birth to about 3 years. It may become noticeable at the time of weaning but in later years may indicate significant anxiety.

Grasp reflex

If the palm of a newborn baby is stroked the fingers will automatically close and the resulting grip may be strong enough to allow the baby to be lifted into an upright position. Conversely if the back of the hand is stroked the hand may open. When the baby sucks, not only will the eyes open but baby may form fists and cross the arms on the chest. This is the protective or clutching reflex. When the baby is feeding the legs may also cross.

Moro or Startle reflex

This reflex can be elicited spontaneously by a sudden noise or if there is a sudden alteration in the baby's position. In its most obvious form the arms are rapidly and symmetrically abducted (thrown outwards) and the hands opened, followed by bringing the hands together and closing them. This reflex will be absent if baby is very hypotonic or very premature. It will be asymmetrical if there has been difficulty in delivering a shoulder resulting in undue traction on an arm causing stretching of the brachial plexus

With traction there may occasionally be an associated fracture of the clavicle (collar bone). This asymmetry due to weakness is called an Erb's paresis and unless there has been complete avulsion of the brachial plexus a full recovery will be made over the next few weeks without any intervention for the paresis or the fracture.

Crossed extension reflex

This reflex when used in conjunction with the placement reflex will allow a reasonably accurate assessment of the baby's gestational age to be made. If baby is placed on the back (supine) and one leg straightened and the sole of that foot stroked, then the opposite leg will take up one of several positions, depending on the gestational age of the baby. If the hip and knee remain flexed, then baby is at 37 weeks gestation or less. If the leg is extended but not adducted (crossed inwards) then gestation is at 38 weeks. If the leg is extended and adducted and brought to the inside aspect of the ankle then baby is at 39 weeks. Finally the leg may cross the shin or be placed on the shin and this would suggest a full term or an overdue baby. The combination of the placement and crossed extension reflexes allows an assessment of the baby's gestational age to be made to within half a week.

Placement or stepping reflex

In carrying out the placement or stepping reflex the back (dorsum) of baby's foot is brought up to touch the under edge of a table ledge while the baby is held vertically. Automatically the foot will then be placed on the table surface in a manner dependent on baby's gestational age. The foot may be placed with the toes first (37 weeks gestation or earlier), or the flat part of the foot (38 weeks) or the heels placed first (full term). After several attempts it will also be seen that most babies place the right foot first. Babies that consistently place the left foot first will usually become left handed and often a family history of left handedness can be elicited.

Oculo-vertical reflex

If after the full examination when most babies are upset, holding the baby in the vertical position by sitting baby in the palm of the hand and supporting the front of the chest with the other hand, within a few seconds most babies will open the eyes and sit quietly provided their temperament is equable. If on the other hand baby continues to cry or be unsettled for several minutes then the temperament will be demanding or fractious. This reflex indicates the baby's basic nature but of course how baby is managed (nurture) which is in the future will be influential. Most babies have an equable temperament in the first few days when the oculo-vertical reflex is performed. Remember a baby with even the most equable temperament will be difficult if colic with paroxysmal abdominal pain or silent gastro-oesophageal reflux with oesophagitis and heartburn develops

Laterality

Approximately 90% of people are right handed. It has been suggested that this is because in utero the right upper limb of the foetus receives blood from a vessel that arises more proximally and earlier from the aorta than that which supplies the left limb. This therefore results in earlier development of the right arm, hand and thumb.

All foetuses suck the thumb during pregnancy with 90% preferring to suck the right thumb. The majority of these babies will have the head turned to the right side while in the smaller group of left thumb suckers the head will be turned to the left. Thus there is a preference for turning the head to the side of the thumb preferred for sucking. Right thumb suckers with the head turned to the right will become right handed later, while left thumb suckers with the head turned to the left will become left handed. Body position in utero has no bearing on the thumb to be sucked.

Chapter 4 : The first week

Nuchal Folds

Fluid collections (oedema) will accumulate in dependent areas of the body. During the first trimester of pregnancy fluid will collect in the neck of the foetus because the foetus is lying on its back and also because the skin in the neck is very lax. Between 10-14 weeks gestation when the foetal lymphatic system is developing there is an opportunity to detect abnormal fluid collections. These can be seen as a nuchal fold translucency and are able to be measured on ultrasound scanning. The more fluid that has accumulated then the greater the risk that a foetal abnormality may be present. After 14 weeks the foetal lymphatic system has developed sufficiently enough to drain away any excess fluid.

Increased fluid accumulation as measured by nuchal fold thickening may mean a foetal cardiac abnormality, a chromosomal abnormality (Down syndrome, Turner syndrome), or that the foetus has failed to move as in a neuro-muscular disorder. In Down syndrome the gene for type 6 collagen is over expressed resulting in connective tissue that is more elastic.The reliability or sensitivity of the ultrasound scan will depend on what standard deviation above the normal range is chosen for measuring nuchal fold thickening.

Amnionic bands (Constriction bands: Streeter's bands)

These constriction bands are present at birth and are very uncommon occurring in about one in several thousand births. They range from a shallow circular constriction in the skin of an extremity, usually the lower third of the leg, down to a deep constriction

Chapter 4 : The first week

extending to the bone. If the constriction is deep enough the arterial supply to the limb or digits is impaired leading to deformity or amputation. The fingers, forearms, toes and legs are most commonly involved in deformities or amputations. When an amputation occurs there is no associated growth retardation or bony defect of the proximal parts of the limb. Amputated extremities may be found at delivery if there has not been enough time for absorption of the amputated parts to take place.

The cause of constriction bands is uncertain. It is most likely that for no apparent reason there is early rupture of the inner amnionic sac without any damage to the outer chorionic sac. The outer surface of the ruptured amnion produces numerous fibrous string like tendrils. The inner exposed part of the chorion has a similar but less marked tendency.

The fibrous strings and the free edge of the ruptured amnion allow for entanglement of the foetal extremities. In addition the foetus may become attached to the exposed chorion and sustain superficial abrasions to the skull, sacrum, knees or elbows. Fibrous strings can be seen attached to the abraded parts or if they encircle fingers or toes soft tissue fusion of the digits will occur with or without deformation. The fused and usually deformed digits have fibrous strings attached to them also.

Note that syndactyly (soft tissue fusion) due to constriction bands involves the distal portion of the digits, as compared with familial soft tissue syndactyly where fusion is seen at the base of the digits and there is no deformation.

Orbicularis oris hypoplasia

When some babies cry the mouth will seem to be asymmetrical. This is because there is relative hypoplasia of one of the lower quadrant muscles in the orbicularis oris group. The muscles in the upper and lower quadrants are responsible for mouth movement and shape. This finding is more common in the left lower quadrant and is less likely to be seen on the right side. The asymmetry is more noticeable with crying and tends to become less obvious with age.

There are two types of orbicularis oris hypoplasia; a transient form predominantly on the left side and which is not an index of any other abnormality and a persistent form usually on the right side and which can be associated with other abnormalities such as congenital hearing defects. Orbicularis oris hypoplasia will mimic a facial palsy but these babies have an intact facial nerve.

Facial nerve palsy

The significant risk factors for an acquired facial nerve palsy in babies are forceps delivery, a birth weight greater than 3.5kgm and in first babies. Trauma occurs to the facial nerve as it emerges from the skull at the stylomastoid foramen where it is subjected to pressure from the forceps blade at this point of egress. It can occur in some babies not delivered with forceps and then is usually due to pressure on the facial nerve by the maternal sacrum during labour. With a facial palsy the cheek on the affected side will be smooth and the mouth on that side will not move or move only slightly. The forehead on the same side will not wrinkle with crying and the eyelids will not close. The eye should be kept moist with a wetting solution such as methylcellulose drops and the eye may be taped closed if necessary to prevent corneal dryness. Facial palsies usually recover completely but a severe case may take several months.

Brachial plexus injury (Erb's and Klumpke's palsy)

The brachial plexus which is responsible for the nerve supply to the arm is formed by the lower four cervical nerves (C5,C6,C7,C8) and the first thoracic nerve (T1). Injury to the upper part of the plexus (C5&C6) is known as an Erb's paresis or palsy and this makes up about 75% of brachial plexus injuries. Lesions affecting C5,C6,C7 occur in about 20% while the remaining 5% involve C8 &T1 (Klumpke's palsy).

The two factors most likely to be associated with a brachial plexus injury are excessive weight of the baby or a difficult delivery. In most cases there is stretching of the upper nerves of the plexus due to traction on the shoulder during the delivery of the after coming head as in a breech delivery or to turning the head away from the shoulder during a difficult vaginal delivery. Injury to the lower plexus nerves is more likely with traction on the trunk in a breech presentation or with traction on an abducted forearm during a normal delivery

The most common findings in a brachial plexus injury are haemorrhage and/or oedema within the nerve sheaths of the brachial plexus.

Avulsion of the roots of the plexus from the spinal cord or tearing of the actual nerves is rare but can occur if traction is severe enough. A fracture of the clavicle or the upper humerus with subluxation (displacement) of the shoulder may also occur in a brachial plexus injury. In milder forms of brachial plexus paresis the Moro (startle) reflex will be asymmetrical with limitation of movement on the affected side. In more severe forms the affected arm will hang limply and be adducted and internally rotated with extension of the elbow, pronation of the forearm, flexion of the wrist but with preservation of the grasp reflex. In the lower plexus injuries the intrinsic hand muscles are affected and the grasp reflex will be absent. There may in that case be an associated Horner's syndrome with a constricted pupil (miosis) and lowering of the upper eyelid (ptosis) on the same side as the palsy. If Horner's syndrome does not resolve the iris will fail to become pigmented and heterochromia iridum (irises with different colour) develops. Rarely there can be involvement of C4 affecting the diaphragm and leading to respiratory distress. A chest X-ray will show elevation (eventration) of the hemi-diaphragm. The majority of brachial plexus pareses recover completely with signs of recovery becoming apparent in the first two weeks and with full recovery of function before two months. The usual brachial plexus paresis does not require any more than a range of passive movement exercises commencing about a fortnight after the delivery. Minor cases do not need any treatment.

The Skin

The skin consists of three main layers – the epidermis, the subcutaneous tissue and the dermis. The outer layer of the epidermis, which is the protective layer (stratum corneum) is poorly developed in the newborn baby. The subcutaneous layer contains little fat and the more premature the baby then the less fat that will be present.

Therefore the skin will be thin and taut and the veins will be clearly visible. In a baby where there has been placental insufficiency the fat stores will have been mobilized and the skin will appear loose and wrinkled. The normal full term baby shows little or no skin desquamation (peeling) until 24 - 48 hours after birth. A premature baby may not show desquamation for 2 – 3 weeks after birth. Desquamation at birth indicates placental insufficiency and as mentioned the skin is thin, loose and wrinkled. In a normally nourished baby if the skin is desquamating at birth then there will have been an episode of acute intra-uterine hypoxia.Therefore the maturity and nutritional state of the newborn baby are reflected in the keratinization process when the skin desquamates to reveal normal underlying skin. Another cutaneous sign of maturity in the newborn appears in the circulatory state of the baby. In the premature baby there will be an exaggeration of acrocyanosis (blue hands and feet) and cutis marmorata (marbling) of the skin when a lacelike pattern of normal blood vessels will be seen in response to cooling. These vasomotor findings are seen in the normal term baby but are not as exaggerated and are of shorter duration.

Sebaceous gland hyperplasia

These are close set yellow-white papules which are seen on the nose in all babies but also on the upper cheeks, ear lobes and chin. Maternal androgens (male hormones) from the ovaries and adrenal glands cause fat gland hypertrophy.

The papules become yellower if the infant is jaundiced probably due to an increased uptake of fat soluble bilirubin. The speckled appearance to the nose will last up to 6 weeks and then disappear, returning again at puberty.

Milia

These are called milk spots and are apparent as firm milk-white papules either singly or in crops. They are commonly seen on the face, forehead, ears and chin but not the nose. They are retention cysts of the pilo-sebaceous follicles and appear as inverted epidermal inclusion cysts which enclose minute keratin pearls which can be extruded by gentle pressure. They are seen in about 50% of babies and disappear after a few weeks.

Miliaria

These are small clear blebs or vesicles which are set close together and easily ruptured and which result from sweat retention in the immediate neo-natal period. They are not seen at birth but can become apparent in a few hours. They are typically seen on the forehead but may also be present on the chest and arms. In the days before the air conditioning of nurseries, miliaria were very common. They are quite harmless and rupture in the first few days. Treatment is conservative and consists in the wearing of light clothes and placing baby in a cooler and less humid environment. There are two types of miliaria:

1. Miliaria crystallina which are clear and occur when sweat is trapped beneath the stratum corneum.
2. Miliaria rubra where sweat is trapped deeper down in the dermis and may set up an inflammatory reaction. The vesicles will then appear reddened and have a pustular component. Unlike miliaria crystallina which appear on the first day, miliaria rubra are not seen until after the first week of life.

Toxic erythema of the newborn: (flea bite dermatitis)

The typical lesion consists of a pale central slightly raised papule surrounded by a flat poorly demarcated reddened macule. They look like heat spots or flea bites and can occur anywhere on the body except the palms and the soles. Typically they appear on the face, buttocks, trunk and limbs. They range from a few discrete spots to dozens and often they coalesce. They are not present at birth but usually appear on the second day although they are occasionally seen in the first 24 hours. The rash comes and goes and may change its site in a matter of minutes. The lesions develop over a few days but as a rule they have disappeared by the second week. Toxic erythema of the newborn is seen in about 30% of babies and there is said to be no cause. My impression however is that this rash is seen only in babies where there is an atopic family history in one or both parents. ie: asthma, hayfever or eczema. If a biopsy is taken (and this is not necessary) the lesions (both papules and macules) contain eosinophils. They do not upset baby, are not influenced by any treatment and disappear spontaneously.

Transient Neonatal Pustular melanosis

This condition unlike miliaria and toxic erythema is present at birth. These large blisters evolve to become large flaccid pustules. They are seen mainly on the trunk and buttocks but can appear anywhere on the body and even the palms and soles. They rupture within the first 48 hours forming a brown crust or collarette which then separates leaving clear or slightly pigmented skin. The condition is benign and has no known cause. It needs no treatment and is uncommon in Caucasian babies

Sucking blisters

These are seen in healthy term babies and are caused by the unborn baby sucking the wrists or fingers and occasionally the feet during the pregnancy. They can be seen in rows along the wrists or hands and appear in varying stages of resolution, ranging from a faint outline only to an obvious blister. They are usually oval in shape, disappear fairly quickly and no new blisters appear after birth unless baby continues to suck the fingers or wrist

Cutis aplasia

This represents the congenital absence of the epidermis, dermis and sometimes the subcutaneous tissue. The lesions are round and clean and may be quite deep and have a punched out appearance. They may be covered by a thin shiny translucent membrane. They are usually seen on the vertex of the scalp and are usually single but can be multiple and have a symmetrical appearance. With the evolvement of an epithelial covering a smooth scar develops which is devoid of any adnexal structures and thus no hair grows later on the aplastic surface.

Ischaemic ulcer

The ischaemic ulcer is due to pressure necrosis and can be difficult to distinguish from cutis aplasia. Ischaemic ulcers result from uterine contractions pressing the affected part usually the skull against the maternal pelvis. They are mainly seen over the parietal eminences of the skull and thus are more laterally situated than the more midline cutis aplasia defect. They may not necessarily be present at birth and are not always round or oval but can develop within hours of birth becoming red and swollen before ulcerating. The ulcer has undermined borders and heals slowly. Once healing has occurred hair will grow.

Caput succadeneum

In every vaginal delivery there is some degree of trauma to the presenting part. The term caput applies to the oedema and bleeding of the presenting part and is apparent as a boggy swelling with bruising in the connective tissue layer of the scalp. Occasionally a friction blister or ulcer may be seen over the caput. The caput resolves spontaneously within a few days.

Cephalhaematoma

The scalp is composed of five layers – skin, connective tissue, galea aponeurotica, loose areolar tissue and the innermost layer the periosteum

A caput occurs in the connective tissue layer and can cross the mid-line. The uncommon sub-galeal haematoma occurs in the loose areolar tissue below the galeal aponeurotica. It is not restricted by the periosteum which it is above and can cross the midline and even spread to the nape of the neck or the lower orbital ridges.

A cephalhaematoma lies beneath the periosteum but outside the skull. It is therefore bound down by the periosteal membrane and is restricted by the suture lines to which the periosteum is attached. Cephalhaematomas are therefore found on either side of the midline of the skull in the parietal area and do not cross the midline. In the uncommon case of an occipital cephalhaematoma however, they can cross the midline. Cephalhaematomas are usually unilateral but can be bilateral. They are present after birth as a soft fluctuant blood filled swelling which later calcifies becoming crackly around the edges, and then becoming hard as ossification occurs. The bony mass is gradually re-modelled back into the shape of the skull. The whole process may take from a few weeks up to several months depending on the initial size of the cephalhaematoma. Cephalhaematomas are common after a vacuum extraction but can be present if there has been excessive moulding of the skull during presentation and passage through the birth canal.

A very large cephalhaematoma in a male infant should elicit questioning regarding a family history of a bleeding disorder such as Haemophilia and if necessary arrangements made for an urgent factor VIII assay. Again in either a male or female infant with a large cephalhaematoma the diagnosis of von Willebrand's disease should also be considered.

As the red cells in the cephalhaematoma break down the bi-product of red cell destruction (bilirubin) will increase and contribute to jaundice. Cephalhaematomas should be left alone and certainly not needled and aspirated since they fill again and infection can follow.

Chapter 5 : Birthmarks and Skin Rashes

There are two types of vascular birthmarks. Haemangiomas and vascular malformations.

Haemangiomas

These show endothelial (lining of blood vessel) proliferation and behave like neoplasms (tumours) in that they exhibit acceleration in growth. They are not always present at birth.

Haemangiomas may be superficial (strawberry birth mark), deep seated (cavernous haemangioma) or a combination of superficial and deep seated (mixed). They are characterized by an initial proliferative and a later involutional phase and will eventually disappear.

Strawberry naevi (superficial haemangiomas)

They are usually absent at birth although a precursor in the form of a white oval macule (flat mark) traversed by a few veins may be noted at that time. They then become red and elevated and grow steadily during the first 6 months, then slow down and begin to involute at about 1 year of age. The typical lesion is a raised rough surfaced swelling which resembles the outside of a strawberry. Resolution may take years depending on the size of the haemangioma. By 5 years 50% have gone, 70% by 7 years and 90% by 9 years. During regression the birth mark changes from a bright red "strawberry" to a dull purple with central white or grey areas where fibrosis due to atrophy of the dilated blood vessel begins. The regression begins centrally and extends peripherally. In about 20% of cases hypopigmentation, atrophy, scarring or telangiectases may persist. If the lesion was very large redundant skin will be left.

Chapter 5 : Birthmarks and Skin Rashes 51

They are seen in about 2.5% of babies within 1 month and about 10% by 1 year of age. They are three times more common in females. They are most commonly seen on the head and neck (60%), the trunk (25%) and the extremities (15%). Usually they are of cosmetic significance only but complications can include bleeding, ulceration and infection. Areas of friction such as the perineum or the lip or when interference to the airways occurs as when they are beneath the tongue or in the neck may mean some form of intervention is needed. If the haemangioma is near the eye vision may be obstructed and with compression of the globe, glaucoma (increased pressure in the eyeball) could develop. If the strawberry naevus is in the midline in the lumbo-sacaral area then MRI of the spine is needed to exclude an underlying tethered cord.

Treatment options for strawberry naevi include any of the following

- observation only.
- compression of the birth mark which may take years and cause anxiety and psychological disturbance.
- systemic corticosteroids when if there is going to be a response it will be seen by regression within 2 weeks. Steroids are more likely to be effective if the naevus is growing rapidly.
- intralesional steroids which give a rapid response but can cause a haematoma.
- surgical excision which will lead to residual scarring
- cryotherapy (liquid Nitrogen) which could cause residual scarring and hypopigmentation.
- laser therapy which depending on the wave length can cause residual scarring.
- interferon which blocks endothelial motility and proliferation and is useful if steroids have failed. It is not to be used in conjunction with steroids.

Cavernous haemangioma (deep seated haemangioma)

They may be present at birth, and can grow very rapidly and usually have a poorly defined border. The overlying skin is normal but has a bluish discolouration since the cavernous haemangioma is composed mainly of large venous channels. They have a cystic lumpy feeling and with crying become larger and a darker blue colour. In time they slowly disappear.

Vascular malformations (telangiectatic naevi)

These have normal endothelial turnover and are structural anomalies that present at birth. They are therefore truly congenital and grow commensurately without acceleration in size.

There are two divisions.

1. Simple naevus (Salmon patches, angel's kiss, stork bite, naevus flammeus neonatorum) and all resolve spontaneously.

2. Port wine stain (naevus flammeus) which does not resolve spontaneously.

Simple naevi

These are red macules (flat) which have an indistinct border, tend to blanch on pressure and do not have a dermatome distribution. They are seen in about 40% of newborn babies. The usual sites are the eyelids (45%), forehead (angel's kiss) or glabellum (33%), nuchal area (nape of the neck – stork bite) (80%) and mid occipital area in about 10%. They can also be seen on the nose and upper lip. Most resolve within the first year or two depending on size but about 10% in the nuchal area remain (Unna's naevus).

Port Wine stain

These are seen in about 2-3/1000 babies. They are composed of mature dilated capillaries lying just beneath the epidermis. They are usually unilateral and seen on the face but can occur anywhere. They are purple to red in colour, have a dermatome distribution, a well-demarcated border and do not blanch with pressure. With time the colour becomes darker

If the ophthalmic division (first branch) of the 5th cranial nerve (trigeminal nerve) is involved then the infant will develop Sturge Weber syndrome and the capillaries of the meninges, the skin and orbit will also be affected. Treatment with a yellow dye 585nm flash light pulsed laser that specifically destroys blood vessels without damaging surrounding tissue may result in complete resolution.

Other Birth Marks

Mongolian Spots or patches   (Baltz Spot, Melanocytic naevi):

These are flat usually well demarcated birth marks made up of collections of melanin producing cells located deep within the dermis. This pigment absorbs all colours except blue and consequently the birth marks have a blue or blue-grey appearance. They can be seen in any ethnic group but are more likely when the degree of natural pigmentation is increased. They are rarely if ever seen on the face, the abdomen, the palms or the soles. The most common site is the lumbo-sacral area or the buttocks. They gradually disappear over the first 2-3 years. Sometimes there is incomplete resolution in the sacral area. They are present in about 10% of Caucasian infants, 95% of Negro and Polynesian infants and about 80% of Asian babies. They are not to be confused with bruises secondary to child abuse and have in the past been called Social worker's bruise.

The first recorded description of Mongolian spots was by a new world missionary Father Gumilla in 1745 who noted that in Indian babies they were present at birth and gradually faded as the infant acquired its natural colour. It has been suggested that Mongolian spots originated in Negroes and spread from them to the Mongol and later the Caucasian races by the invasion from Asia of the Huns and Mongols. - hence the name.

Café au lait patch (Hyperpigmented macule)

These can be easily missed in the newborn since they appear as pale tan coloured flat marks. They are variable in size and shape and can be found on the face, trunk or limbs. They are not always present at birth.The shade of pigmentation is appropriate to the colour of the background skin, the colour is uniform and the edges of the macule are distinct. Small single birth marks <2cms are seen in about 20% of children normally. Much larger irregular birth marks usually on the trunk raise the possibility of McCune Albright (precocious puberty) syndrome, while the presence of 6 or more café au lait patches >0.5cms diameter in a baby indicates the possibility of neurofibromatosis, a variable condition with several sub-types including 8th nerve tumours, cafe au lait patches and Lisch nodules in the iris.

Epidermal naevi (warty naevus; sebaceous naevus)

These birth marks represent excessive development of the epidermal cells and are made up of sebaceous glands, sweat glands, abnormal hair follicles and hyperplastic epidermis. They appear as discrete yellowish cobblestone like lesions which are usually small and oval or linear in shape but in some cases can be quite large. They tend to get a little smaller after birth but usually do not disappear. Unless associated with other abnormalities they have no significance other than as a cosmetic blemish. They become raised and warty in adolescence and should probably be removed at that time. About 20% develop a basal cell carcinoma in them in adult life.

On the scalp and face they appear yellowish because of a prominent sebaceous gland component and are called sebaceous naevi. On the abdomen, chest or limbs they are less common and are called warty naevi.

Skin rashes

Neonatal Acne

This is a self limited condition which will persist until maternal androgens which have been acquired trans-placentally have dissipated. These acne-like lesions are not present at birth but appear in the first few weeks. They occur in crops on the cheeks, nose and chin and occasionally on the forehead. It is more common in male infants. Infants with neonatal acne will tend to develop severe acne later in life. This is because the pilo-sebaceous unit is hyper-responsive to androgens from mother and which have crossed the placental barrier. They are the result of a genetically determined end organ response.

Infantile acne

In infantile acne which first occurs after 3 months there is only a small preponderance of males over females. There is a strong genetic component and a familial tendency towards acne is usually present.

Cradle Cap (seborrheic dermatitis)

Cradle cap appears in the first weeks of life as a scaly flaking non-itchy crust over the scalp, eyebrows and behind the ears. Later it may become more widespread and can affect the skin flexures (intertrigo) and the perineum. Heat, moisture and sweat retention combine to cause maceration of the skin folds (seborrheic dermatitis). Cradle cap is usually beginning to resolve at about 4 months of age. It may be confused with atopic dermatitis (eczema) but this condition does not appear until about 3 months and is itchy. The atopic infant will be itchy uncomfortable and unhappy while the seborrhoeic infant is comfortable and happy.

Seborrhoeic dermatitis may be complicated by secondary infection with Candida (thrush) especially in the napkin area and in that case satellite lesions which are reddened and circular with a scaly peripheral border extend beyond the well demarcated red areas of seborrhoeic dermatitis. If untreated these lesions coalesce. Living Candida does not penetrate living skin and the conspicuous skin damage seen is due to irritant yeast products and toxins which filter into the inflamed skin after the Candida organisms die and disintegrate, thus making culture difficult.

Cetrimide cream applied thinly to the scalp, eyebrows and behind the ears and then washed off shortly afterwards in the bath is effective for cradle cap. For flexural intertrigo skin surfaces should be dry and Zinc and Castor oil ointment applied to the flexures frequently. Castor oil is a vegetable oil and acts as a barrier while Zinc is mildly antiseptic. If thrush is also present then an anti-fungal cream will also be needed.

Napkin rashes

The best looking skin one will ever have is when one is a newborn baby. Signs of skin irritation can develop in the first month or even earlier by the end of the first week when flaky skin may appear. In utero the skin is protected by a lipid rich film (vernix caseosa) which disappears after birth but can also be rubbed off.

Factors influencing the appearance of napkin rashes include skin wetness when excessive hydration makes the skin more susceptible to the effects of friction forces leading to erosions and maceration. Wetness also increases skin permeability and increased wetness in the napkin area will facilitate rapid growth of micro-organisms. Bowel motions contain enzymes such as proteases and lipases which in themselves can be skin irritants. These enzymes are activated at higher pH levels which result with the interaction of bowel motions, urine and bacteria. Bacteria in the bowel motions produce ureases and these enzymes convert urea in the urine to ammonia which also contributes to raising the pH and allowing activation of irritating proteases and lipases. Breast fed babies have a lower incidence of moderate to severe napkin rashes because the faeces of breast fed babies are acidic and therefore there is reduced urease, lipase and protease activity. Candida albicans (thrush) is usually a secondary invader of damaged skin.

Whether cloth or disposable napkins are used frequent changing of the napkins will help prevent napkin rashes. Cloth napkins should be machine washed and rinsed to remove any detergent or anti-bacterial solution. They will also be softer and less likely to irritate the skin if they have been tumble dried for a few minutes. Disposable napkins which have an absorbent polymer gel provide better protection because they draw moisture away from the skin and keep it dry. However there is still an occlusive outer plastic layer that raises the humidity of the skin. This warm, humid and occlusive environment beneath the napkin is ideal in encouraging bacterial growth and thrush. The newer Gortex materials are permeable to air and vapour but impervious to liquid. Placement of such a microporous film as a napkin cover will significantly reduce humidity and wetness of the skin. Disposable napkins take ages to decompose in landfill areas.

Contact Dermatitis

If Napkin rash is confined to the convexities such as the buttocks, inner aspect of the thighs, the pubic area and the scrotum and since these are the areas in intimate contact with the napkins then the resulting rash is called contact dermatitis. The skin creases will be spared. Contact dermatitis is usually due to ammonia produced by micro-organisms in the bowel motions or residual irritant detergents which persist if the napkins have not been adequately cleaned. This irritant dermatitis can become secondarily infected with bacteria resulting in pustules, nodules or skin erosions. Ammoniacal contact dermatitis is not usually seen before three months of age.

Tidemark dermatitis

This is a band of erythema (redness) seen on the skin at the margins of the napkins and is due to contact with the elastic band or the edge of the napkins.

Granuloma gluteale infantum

This is a rash consisting of reddish-blue nodules of varying size from a few millimeters to centimeters and is preceded by napkin dermatitis. It is limited to the gluteal area (buttocks), inner part of the thighs and the pubic area. It is caused by prolonged application of potent fluorinated steroids to the inflamed skin

The nodules are well demarcated and resolve over a few weeks and with or without treatment but they leave a flat brown patch.

Peri-anal erythema is due to the action of acidic or alkaline loose bowel motions on the skin in either the breast or formula fed baby.

Flexural intertrigo (seborrhoeic dermatitis)

This rash is seen in the skin creases in the napkin area where moist skin surfaces rub together causing maceration and inflammation. The skin convexities are spared. Flexural intertrigo can be complicated by a supervening thrush (candida) infection. When thrush appears circular reddened and flattened lesions with a flaky border are seen and these eventually coalesce unless treated with an appropriate cream. Flexural intertrigo is best treated at each napkin change with a barrier cream such as Zinc and Castor oil ointment where the Castor oil (vegetable oil) component prevents the rubbing together of adjacent skin surfaces and the Zinc acts as a mild antiseptic. Remember that some babies skin reacts to a mineral oil ointment such as Vaseline.

Eczema (Atopic dermatitis)

Eczema is seen mainly in infants and children and is also usually associated with a family history of other allergies. About 10%-15% of children will have had eczema. True eczema is rarely seen before four weeks of age and usually has declared itself by three months. Of infants who will get eczema 75% will have signs by six months. The earliest presentation may merely be a little dryness or roughness of the skin and this may be noted after the first week of life. It is a symmetrical condition seen initially over the face, trunk and limbs but sparing the palms, soles and napkin area. Involvement of the trunk and limbs takes the form of large poorly defined scaly areas or discrete coin shaped (nummular) lesions. Facial involvement is most prominent on the cheeks and is symmetrical. The forehead and chin can also be affected but the nose and peri-nasal skin folds are not. After about one or two years the distribution of eczema changes, clearing from the face and now beginning to affect the skin flexures (elbows and knees) wrists and ankles again in a symmetrical manner. Eczema is associated with dry skin and follicular hyperkeratosis with obstruction of the sweat ducts, this combination of factors promoting itching which of course is a feature of eczema.

Another associated finding is keratosis pilaris with keratin plugging of the hair follicles giving a plucked chicken appearance to the skin and which is easily seen over the outer aspect of the upper arms. Infants with eczema should be protected from itch producing situations such as excessive perspiration due to over heating at night or excessive clothing and contact with wool. Other irritants can include synthetic fibres, heat, soap and shampoos. Prolonged or frequent bathing will cause skin dryness leading to itching. Clothing should be soft and made of cotton or a cotton polyester. Acrylics, nylon and wool should not be in direct with the skin.

A dry skin means that the stratum corneum (epidermis) layer is dry and this layer of skin in order to function properly must contain an adequate water content. A hydrated stratum corneum acts as a barrier to the penetration of irritants, bacteria and toxins and also as a membrane helping to retain body fluids. Soaking in water will hydrate the stratum corneum and the infant will feel more comfortable while in the bath. However after the bath the skin becomes drier than before because the lipid film which holds water in the stratum corneum has been removed.

Breast feeding may reduce the risk of eczema if mother eliminates major allergens from her diet. If formula fed, baby is better on a hypo-allergenic formula rather than soy or cow's milk and delaying the introduction of solids until six months can also be helpful.

In the bath avoid soap or chlorinated water and use a water dispersible bath oil or a soap substitute such as emulsifying ointment. Apply a skin emollient such as Cetaphil or Lipobase immediately after the bath as these emollients are soothing, protective, greaseless and help keep water in the stratum corneum.

Any steroid applied to the skin will be absorbed to some extent and generally ointments promote more absorption than creams. Fluorinated steroid ointments or creams should not be applied to the face of babies. Children up to four years of age should not be treated with steroids applied to the skin for periods longer than three weeks at a time. Almost all infants with eczema are colonized with Staphylococcus aureus and if there is any suggestion of a secondary skin infection, then the appropriate antibiotic should be given by mouth for at least ten days. Night time itching and scratching can be helped with anti-itch antihistamines such as Vallergan syrup given before bed. Cotton mittens and short nails are also helpful.

Chapter 6 : The Head

The Fontanels

Fontanelle is the French diminutive for fontaine meaning fountain. Fontanels are unossified membranous intervals felt as a depression at the angles of the parietal bones of the skull. The anterior and posterior fontanels are situated in the longitudinal mid-line of the skull. The anterior fontanelle is larger and lies at the junction of the sagittal, coronal and frontal sutures (junction of parietal and frontal bones). There is a wide variability in shape but no significant difference between the mean length and width. The size at birth varies between 0.6 -3.6cms with a mean of 2.1cms. The anterior fontanelle closes by about 18 months – 2 years of age. Closure of the anterior fontanelle will be seen in 1% of full term babies by 3 months, about 40% by 12 months and 95% by 24 months. The median age of closure in full term infants is about 14-15 months. The posterior fontanelle is triangular in shape and is situated at the junction of the sagittal and lambdoid sutures (junction of parietal and occipital bones). The posterior fontanelle is small (less than 0.5cms) at birth and has closed by three months.

The skull grows rapidly from birth to 7 years with the greatest increase being in the first year this reflecting rapid brain growth during that time. The head circumference for full term male infants at birth ranges from 33-37.5cms and for full term females from 32.5-37cms, these measurements lying between the 3rd-97th percentiles. As a rule of thumb guide the head circumference grows 2cms/month for the first 3 months, 1cm/month for the next 3 months and 1/2cm/month for the final 6 months of the 1st year of life. ie: about 12cms growth in head circumference in the 1st year. The presence of the fontanels allows for the initial rapid period of growth of the brain and after the fontanels have closed further growth will continue in the fibrous tissue of the sutural ligaments.

A head circumference below the 3rd percentile in the absence of overriding sutures is consistent with microcephaly but not necessarily mental retardation. Primary microcephaly is caused by anomalous development of the foetal brain in the first 7 months of gestation. It is usually due to an autosomal recessive gene, or one of a variety of chromosomal disorders or else an intra-uterine infection with rubella, cytomegalovirus or toxoplasmosis. Other possible causes are maternal hyperphenylalaninaemia with maternal serum phenylalanine levels over 15mgm/100mls during the pregnancy or maternal irradiation early in pregnancy. In true (primary) microcephaly the anterior fontanelle is small or absent and the forehead narrow and receding.

Secondary microcephaly is the result of an insult such as infection, trauma, metabolic disturbance or anoxia occurring after the 7th month of gestation. In these instances the head circumference will be in the normal range at birth but subsequently over the next few months there will be only a slow increase in head circumference growth.

Small fontanels with overriding sutures are of little significance and are usually due to intra-uterine compression of the skull.

The anterior fontanelle is best assessed with the infant held upright and quiet and the fontanelle should then feel soft and slightly depressed. Pulsations that can be seen are normal

A large full fontanelle with separated sutures (particularly the squamous temporo-parietal suture) will suggest raised intra cranial pressure. Similar findings of a large fontanelle without raised pressure suggest delayed skeletal growth as in hypothyroidism, bony dysplasias (achondroplasia, cleidocranial dysostosis, osteogenesis imperfecta), vitamin D deficiency or some of the chromosomal trisomies. Accelerated skeletal maturation as in hyperthyroidism will be associated with a small anterior fontanelle.

The so-called 3rd fontanelle which lies about 2cms in front of the posterior fontanelle is due to a parietal bony defect (delayed ossification) along the sagittal suture. It is not a true fontanelle because it does not lie at the junction of the parietal and adjacent bones. It is more likely to be present in Down syndrome, congenital rubella and in low birth weight (2.0kgm-2.5kgm) babies. The metopic fontanelle represents an extremely long extension of the anterior fontanelle which in the process of suture closure has become separated from the anterior fontanelle. Again it is not a true fontanelle but it can be associated with Down syndrome, congenital rubella or cleft lip and/or palate.

Skewed and flat heads

(plagiocephaly and brachycephaly)

After the delivery baby's head may be seen to be flattened on one side or slightly skewed. These changes are positional in nature and due to intra-uterine compression. They resolve spontaneously over a matter of weeks. The head may also be drawn to one side because the large neck muscle (sternomastoid muscle) is tight or short and this will prevent the baby looking to the opposite side. If there has been excessive traction delivering the head some sternomastoid muscle fibres can be torn and in that case after a week or two a lump will become palpable in the muscle – sternomastoid "tumour". The "tumour" is a collection of blood from the torn muscle fibres and as this becomes organized into scar tissue the lump disappears and the resulting scar tissue will contract and draw the head to the affected side – torticollis. This particular torticollis is obligatory because the tight and shortened sternomastoid muscle prevents active rotation and lateral movement of the neck. Gradually plagiocephaly and facial asymmetry will become apparent. In that case physiotherapy with a series of passive stretching and rotational movements will be necessary.

However rather more likely over the next months and largely because of the "back to sleep" campaign where babies are slept in the supine position (on their back) and not prone or on the side, the infant's head may become flattened posteriorly (brachycephaly). If the head is turned to one side then the skull becomes skewed (plagiocephaly). These changes in the skull shape occur because the skull is malleable and babies in the course of a 24 hour day, spend 15-18 hours sleeping in the one position. In these cases however spontaneous movement of the head and neck to either side is possible and thus is not obligatory. To lessen this problem baby should be given time on the tummy when awake.

If there is plagiocephaly the cot is rotated through 180 degrees so that all interesting objects are on the side away from the plagiocephaly.

Interesting surroundings for baby include the parent's bed, the doorway through which mother enters baby's room, windows and large highly visible toys. By making this change the baby will be encouraged to turn the head to the side of interest and therefore counteract the skewing effect.

These are all positional and preferential changes but can be exacerbated if there is limited movement due to poor muscle tone, or poor mineralization of the skull bones. It is important also to exclude any suggestion of early sagittal or coronal suture fusion because with early sagittal suture fusion (synostosis) there will be an elongated boat shaped head while early coronal suture fusion will give a flattened head. In both these cases sleeping position has no bearing on the shape of baby's head. Synostosis of skull sutures requires a skull splitting procedure later. Remember that certain ethnic groups have a flattened occiput in any event thus giving a brachycephalic skull which of course is quite normal.

Positional flat (brachycephalic) or skewed (plagiocephalic) heads are seen in about 1 in 7 infants in either mild or obvious degree and all resolve spontaneously over the next 2 years. Seldom if ever with a positional skull deformation will drastic intervention like an orthotic helmet be necessary. Usually all that is needed is rotation of the cot, encouraging baby to look to either side when awake, "tummy time" and if parents are anxious some gentle passive stretching and rotational movements of the neck.

Chapter 7 : The Mouth

Cleft Lip and Palate

Clefting of the lip, lip and palate or palate alone occurs in about 1 in 700 live births. Cleft lip is seen in about 5% of these babies, cleft lip and palate in 50% and cleft palate alone in 45%. Clefting may be bilateral (25%) or unilateral (75%) and then it is slightly more common on the left side. Cleft lip with or without cleft palate is more common in boys while a cleft of the palate alone is twice as common in girls. The majority of clefts are isolated defects usually due to a combination of genetic and environmental factors. Associations have been noted with some anticonvulsants (phenytoin and valproate), steroids, diazepam and maternal smoking. The recurrence risk is about 15% if there is a family history (1st or 2nd degree relative) of clefts. If there is no family history and no associated syndrome then the recurrence risk is about 1-2%. There is a higher prevalence of cleft palate in left handed people and there is a higher (85%) chance of the cleft being on the left side. There is said to be an increased incidence of cancer early in life among children with congenital malformations including cleft lip and palate. Also children born with cleft lip and palate but no other malformations seem to have an increased risk of mortality not only in the 1st year of life, but throughout the three intervals of life (0-1 year, 1-17years, 18-55years).

Usually babies with cleft lip alone or an isolated cleft of the soft palate have little trouble with breast or bottle feeding. In babies with cleft lip the breast will help fill the lip defect and allow the infant to develop the negative intra-oral pressure necessary for successful feeding. A large wide based teat will help do the same. On the other hand babies with cleft lip and palate will have difficulties generating sufficient negative intra-oral pressure and will need to be bottle fed, preferably with expressed breast milk and using a soft easily squeezed bottle and teat which will allow control of milk volume and flow rate. As a general rule a cleft lip is repaired at about 3 months and the palate at about 1 year of age. If possible the operation is performed by a surgeon experienced in this area.The achievement of normal hearing, normal speech, acceptable maxillary and facial growth and a good orthodontic result is the goal of treatment.

Sub mucous cleft

This is not visible but is diagnosed by palpation of the palate when a wide V shaped mid line notch rather than a rounded curve will be felt at the junction of the soft and hard palates. A sub mucous cleft must also be excluded if the uvula is found to be bifid.

Uvula

A bifid uvula can be a marker for a sub mucous cleft palate which is palpable rather than visible. A bifid uvula is revealed as a small notch in the lower end of the uvula and is found in about 2- 4% of babies. A bifid uvula represents the mildest form of cleft of the soft palate. If a sub mucous cleft is present it can cause problems with middle ear disease later on and the tonsils or adenoids should not be removed in order that velo-pharyngeal competency can be preserved and hypernasal speech avoided.

Epstein Pearls

These are small white cystic lesions found in groups in the midline of the palate at the junction of the soft and hard palate. They are seen in almost all newborn babies and they represent trapped epithelial remnants that become keratin filled cysts. They disappear over several months. Similar white cysts can occasionally be seen on the gums and they are then called Bohn nodules.

Congenital Epulis

A congenital epulis is a large pedunculated benign tumour projecting from the upper or lower alveolar ridge (gingival or gum margins). They are more likely to arise from the lower (mandibular) gingival margin than the maxillary ridge and are more common (8-10 times) in females than in males. Large masses should be excised since they interfere with feeding and if excised they do not recur. Small congenital epuli can be left alone and will often regress spontaneously.

They were initially described in the 1870's and are seen only in the newborn.

Ranula (Salivary mucocoele)

A ranula is a fluid filled retention cyst of the salivary glands which lie beneath the tongue. They can be quite small in which case they rupture within a few days to a few weeks. Rarely they can be very large and big enough to interfere with feeding in which case surgical removal is necessary.

Tongue tie (Ankyloglossia)

The frenulum lies beneath the tongue and arises from a thickening of the genio-hypoglossus muscles which meet in the mid-line beneath the tongue to form a vertical fold (frenulum). The tongue is normally short at birth and as the baby grows the tongue becomes longer and thinner towards the tip. The frenulum also lengthens and stretches, thereby increasing the range of movement of the tongue. Eventually the frenulum appears to recede and is placed well behind the tongue tip. The diagnosis of tongue tie is by no means clear cut and there are no universally acceptable criteria or definitions. The diagnosis is therefore a highly subjective one existing mainly in the eye of the beholder and usually depends on the degree of visibility of the frenulum. The frenulum ranges from a thin mucous membrane band to cases where the frenulum is markedly fibrosed and where the tongue may be fused to the floor of the mouth. It is generally accepted that tongue tie might exist if the frenulum is abnormally short and thick or in an older child when the tongue cannot be protruded beyond the lower incisor teeth.

Many mothers ascribe their baby's feeding problems or child's delay in speech acquisition or indistinct speech to tongue tie. This impression is usually wrong. With feeding babies use the intrinsic muscles of the tongue to raise the body of the tongue to the palate and then lower it, thereby altering intra-oral pressure in order to propel milk into the pharynx. Tongue protrusion is not a feature of the infant sucking. Tongue tie does not cause feeding difficulties and the answer to these problems should be sought elsewhere.

Feeding problems in the first year of life should not mean that the frenulum must be cut.

Babies should never have the thin frenulum cut even if when the tongue is protruded puckering of the tip of the tongue can be seen. There can be bleeding from the deep lingual vein in the frenulum and this may also be complicated by infection. The operation in the newborn period is due to ignorance of the normal appearance of the tongue and frenulum at that time. While tongue tie does not cause delay in speech acquisition it can however contribute to difficulties in the rate and range of articulation. Children under 5 years of age should not be operated on and only afterwards if the tongue tie is severe.

The Teeth

Natal and Neonatal Teeth

Natal teeth erupt in the first month of life while neonatal teeth are present at birth. The incidence is about 1/2000 and they may be part of the normal complement of primary (deciduous) teeth or they may be supernumerary. This would then require a radiological diagnosis in order to make the distinction. The most common site is the central incisors of the lower gum. The teeth may be mobile because the roots are not yet completely formed. If the tooth is freely mobile it should be removed to avoid aspiration. The enamel of the tooth is usually dysplastic and about 50% of natal or neonatal teeth are lost in the first 4 months. They do not cause tongue laceration or irritation of the nipples during feeding. A positive family history of neonatal or natal teeth is often obtained.

Teeth eruption

The primary teeth are erupting from about 6 months to 6 years of age. Primary teeth take up to 2 months to reach full eruption while secondary teeth take 3-4 months. Teeth erupt at a rate of about 1-2mm/month.

Old wives tales regarding teething go back to the 2nd century for the earliest recordings – redness, swelling and drooling were noted. All sorts of conditions and illnesses have been ascribed to teething but they are coincidental or possibly due to some of the medicines offered. Nevertheless there may be a slight elevation in the temperature a few days before the eruption and there can be ballooning of the gums with flushing of the cheek. A bright spot (hectic spot) may appear in the centre of the flushed area. Dribbling and tugging of the ear lobe will become more obvious. If the baby is really distressed with teething an antihistamine with mild sedative side effects can be rubbed on the gums and then allowed to be swallowed.

Milk Caries

Milk caries initially affects the upper anterior (usually canine) deciduous teeth and later the posterior teeth. The lower anterior teeth are not affected because with feeding the tongue extends slightly and covers the lower teeth. Milk caries is due to frequent prolonged contact of the teeth with sugars in formula or expressed breast milk. This situation exists with prop feeding when the infant is put down to sleep with a bottle in the mouth. Carbohydrates in fruit juices or other bottled drinks can also be responsible. White spots on the teeth are the earliest sign of milk caries

Before the actual cavities develop the enamel of the teeth becomes de-calcified giving a mottled appearance with the formation of white spots. Babies should not be prop fed and a cup should be introduced by at least 12 months of age unless baby is still breast feeding at that time. Prolonged breast feeding (average duration of 21 months) is associated with an increased risk of milk caries

Various streptococci and lactobacilli are the initiating cause of tooth decay and these bacteria are found in plaque from tooth surfaces of normal or carious teeth.

Intra-oral bacterial plaque increases acid production (lowered pH value) and it is this combination of increased acid and sugar content that produces caries. Sugar causes fermentation with resulting acid production, and the worst sugar is sucrose. A substance that has a low pH but without sugar is unlikely to produce caries because of the buffering effect of saliva. The development of caries therefore requires the combination of acid and sugar in prolonged contact with the teeth. Sucrose containing medicines can also be a problem since some of them contain up to 70% sucrose. Remember that sugar free sweeteners such as Sorbitol can cause acute and chronic diarrhea because in some children Sorbitol is not absorbed by the small bowel. This then leads on to an osmotic diarrhea accompanied by cramping abdominal pain. If the teeth are varnished with a fluoride solution (50mgm/ml) the development of milk caries can be prevented.

Teeth discolouration

Teeth may become discoloured for various reasons. In a baby who has been deeply jaundiced for some time the jaundice pigment (bilirubin) is deposited in the dentin and the enamel in primary teeth giving them a yellow appearance. Later they may become green or blue-green in colour as bilirubin is oxidized to green biliverdin. Fortunately the colour which affects primary teeth fades with time. Osteogenesis imperfecta may present with similar coloured teeth.

Minocycline a semi-synthetic derivative of Tetracycline can cause staining of the permanent teeth (unlike Tetracycline which stains primary teeth). This is a grey-black discolouration and it can also be associated with abnormal pigmentation of the skin, nails, bones, sclera and conjunctivae. Pigmentary changes in the skin resolve but the tooth discolouration does not.

Rifampicin which causes the red man syndrome, with a sunburned appearance, can stain the teeth red.

Iron may stain teeth black and red wine drinkers can develop a greenish tinge to the teeth, with erosion of tooth enamel on the maxillary canines and incisors but despite this they have a lower prevalence of dental caries.

Chapter 8 : The Ears

The external ear is called the auricle or pinna. The outer rim of the ear is the helix and the curved prominence parallel to and in front of the helix, the anti-helix. The small curved flap projecting backwards and opposite the helix is the tragus and the curved prominence opposite the tragus and at the upper surface of the ear lobe, the anti-tragus. There are three depressions in the ear and from top to bottom they are the scaphoid fossa, the triangular fossa and the concha. The folds which separate the fossae are called crura.

In the first few days of life the neonatal ear is soft and malleable because of transiently high levels of maternal oestrogens which increase elasticity. Within three days the oestrogen levels have fallen and the ears then become less elastic and firmer.

An abnormal looking ear is more often associated with a middle ear anomaly than with any other congenital abnormality.

Low set ears

Usually this is a subjective impression but objectively low set ears can be defined. The length of the ear is measured. If a horizontal line is drawn from the inner canthus of the eye across the face to the ear and if less than 10% of the ear is above that line, then the ear is low set.

Small and malformed ears

These are seen in certain syndromes (Goldenhar's and Treacher Collins which in addition to deafness, are associated with eye, eyelid and jaw abnormalities) and hemi-facial microsomia. An abnormally formed ear usually means an associated hearing deficit on that side. Drugs such as Isotretinoin taken during pregnancy can cause ear abnormalities.

Vertical groove in ear lobe

If baby is very large and with a prominent tongue and hemi-hypertrophy of the body and limbs, then a vertical groove on the ear lobe is supportive evidence of Beckwith-Wiedemann syndrome. Problems with low blood sugar levels in the first few days should be anticipated and monitored.

Crumpled or cupped (lop) ears

This appearance is usually due to intra-uterine compression and the ears will slowly assume their natural position and shape. The appearance can be exaggerated if there is decreased cartilage or an increase in elasticity of the ear as in Marfan's syndrome.

Pre-auricular dimples (pits):

This is a dominantly inherited condition and a dimple may be found in front of one or both ears in about 1% of babies. Other than being a family trait it has no other significance.

Pre-auricular skin tags

These tags are situated in front of the ear and hang by a thin or thick pedicle and can be found in about 1% of babies. If the tag hangs by a thin pedicle then it may be ligated and will then fall off in the first few days. On the other hand if there is a thick pedicle then surgical excision is undertaken at about 12 months of age. In any baby where there is a pre-auricular skin tag a formal hearing assessment should be arranged at about 8 weeks of age. However in the absence of any other ear anomaly the chances of an associated hearing loss are unlikely.

Chapter 9 : The Eyes

The eye grows at about the same rate as the brain and by 30 months of age has reached 2/3-3/4 of the adult size. Newborn babies do see and can respond preferentially to certain visual patterns. If shown a card with a human face, a scrambled human face, or a blank card the baby will turn the gaze more often towards the human face suggesting that baby is programmed for this response rather than having to acquire it by recognition. This is a cerebral cortical response and indicates that there is central nervous system integrity.

Normal ocular alignment is established at about 8 weeks. Before then alignment may shift from internal squinting (esotropia), to normal alignment or divergent squinting (exotropia). For the first 6 weeks infants "track" with jerky eye movements but by 3 months of age smooth tracking is present. True blinking in response to an object introduced into the visual field develops between 2-5 months. The visual acuity of the newborn is about a 20th-30th of the normal adult visual acuity and normal adult vision is achieved by 2 years of age. The newborn babies see best those objects which are near them and also those which have high contrast. The retina consists of a central fovea which is composed of cones responsible for detail and colour and the peripheral rods which mediate movement and brightness. There is some uncertainty about baby's response to colour but by 8 weeks babies possess red and green cones but lack blue cones which appear later at 3 months. For the first 1-2 months therefore they may have an insensitivity to blue colours.

Tears are present at birth but are not produced in response to crying for 1-3 months.

The iris continues to develop for the first 6 months of life and the colour at birth which is usually blue-grey for Caucasian babies may darken later as the melanin producing cells begin to produce pigment. The final colour will usually become apparent by about 6-9 months.

Brushfield spots

These were first described by Thomas Brushfield in patients with Down syndrome. They are present in 85-90% of Down syndrome babies and are variable in size (0.1mm-1mm) and may coalesce. They are seen at the junction of the middle and outer thirds of the iris. Here the iris stroma becomes thinner and the spots show up as stromal fibres. Areas of iris hypoplasia are also seen in most Down syndrome children. Brushfield spots are very difficult to see in brown eyed children. They can also be seen in some other chromosomal disorders and occasionally in Klinefelter's syndrome.

Wolfflin nodules

These are small areas of iris hypoplasia and are seen in about 25% of people. They appear as small dots (0.1mm-0.2mm), uniform in shape and arranged in regular formation and are situated at the junction of the outer 4/5ths of the iris. They should not of course be confused with Brushfield spots.

Cataracts

These are opacities in the lens. If the lens is clear then an uninterrupted red reflex is seen when using an ophthalmoscope. Cataracts may be hereditary, be the result of congenital infection or be metabolic in origin.

Hereditary cataracts are present at birth and are either dominantly inherited but with variable penetrance which is the common form or, of recessive inheritance which is uncommon. Hereditary cataracts involve the central area of the lens and may be static or progressive.

Of cataracts due to congenital infection, congenital rubella is the most likely cause, but toxoplasmosis or cytomegalovirus are other possibilities. Cataracts due to congenital rubella may be unilateral or bilateral, partial or total and they are usually present at birth.

Metabolic cataracts are caused by the rare autosomal recessive condition galactosaemia. The accumulation of galactilol leads to hypertonicity of the lens with consequent disruption of the lens fibres. They are not present at birth.

Coloboma

A defect in the iris such as a notch or a hole will result in a tear drop shaped pupil and may be present in one or both eyes. Other colobomas can be present in the retina appearing as a wedge shaped white area. Colobomas of the iris alone do not affect vision but do mean a sensitivity to bright light.

The Naso-lacrimal duct

Each of the two eyelids has a lacrimal punctum and the one most easily seen is in the margin of the lower eyelid at the inner third junction. Both puncta drain into lacrimal canals and then into the lacrimal sac which is at the upper end of the naso-lacrimal canal. The naso-lacrimal canal then drains into the lower part of the back of the nose. Tears are secreted by the lacrimal gland and are carried across the eye to the groove between the lower eyelid and the eyeball and then into the naso-lacrimal canal. Blinking sucks the tears into the lacrimal sac by capillary attraction.

Congenital obstruction of the naso-lacrimal duct is quite common affecting about 5% of newborn babies. It usually affects one side more than the other and most cases resolve spontaneously over weeks or months. Secondary infection can develop with swelling of the eyelids and the inner border of the lower eyelid. If naso-lacrimal obstruction resulting in a perpetually wet or glistening eye does not resolve by 15 months then the duct should be probed under general anaesthesia. However far more common and seen briefly in most babies is swelling of the eyelids as a result of the delivery. Because of increased pressure from the delivery fluid extravasates into the loose tissue of the eyelids leading to blockage of the lacrimal puncta. This prevents drainage of tears and thus causes wet eyes and

Chapter 9 : The Eyes

sometimes a secondary infection of the eyelids. Bacteria grow readily in a wet warm and dark environment and the eyelids may become quite sticky. If the white part of the eye (sclera) remains white and there is no conjunctival inflammation then eye toilet with breast milk, saline solution or tap water is sufficient treatment. Antibiotic ointments are not necessary unless there is conjunctival reddening. The most common causative agents will be chlamydia, staphylococcus aureus, haemophilus influenzae, group B streptococci, listeria or pneumoccocci. An eye swab for culture should then be taken

Chapter 10 : The Chest

Breasts

Breast tissue is present in the newborn female and male infant. The earlier the gestational age, the less breast tissue will be present. Breast tissue is due to maternal oestrogens which have crossed the placenta. The breasts may be quite large and milk can be expressed from the newborn breast (witch's milk). The nipples should not be too widely spaced although this is a subjective judgment. Widely spaced nipples with a shield like chest may indicate Turner syndrome (short sterile female).

Supernumerary nipples (polythelia) or breasts (polymastia)

Polythelia are also called vestigial accessory nipples and are seen in the newborn as a small dimple or puckering of the skin anywhere along the milk ridge which extends obliquely from the armpit (axilla) to the groin (inguinal area) on the other side of the body. A vestigial nipple if present is usually found below and medial to the breast proper. About 1 baby in 200 will be affected and vestigial accessory nipples can be unilateral or bilateral. They usually represent a familial trait inherited in an autosomal dominant manner and they are permanent. Accessory vestigial nipples in male infants can carry a slightly increased risk of kidney abnormalities and testicular cancer which is about 4-5 times higher than in the general male population. This translates into about 5 cases per 10,000 men per year. However in the absence of any other dysmorphic features there is little to be gained by undertaking genito-urinary investigations.

Accessory breast tissue (polymastia) in contrast to accessory nipples can produce milk and they will increase in size in response to hormones, pregnancy or drugs.

The Heart

The heart is divided into four chambers; the right and left atria and the right and left ventricles. The atria receive blood and the ventricles distribute blood. The right and left sides of the heart are separated by partitions called septa. After birth there is no communication between the right and left sides of the heart which now function in parallel. The right atrium and right ventricle receive de-oxygenated blood returning from the body and send it on to the lungs. The left atrium and left ventricle receive oxygenated blood from the lungs which is then distributed around the body.

The foetal circulation

Foetal blood is carried to the placenta by two umbilical arteries and returned to the foetus by the umbilical vein. The umbilical vein eventually becomes a large vessel called the ductus venosus which via the liver drains into the inferior vena cava.

The inferior vena cava then drains into the right atrium where it is directed through a window in the right atrial septum called the foramen ovale into the left atrium

Some blood from the inferior vena cava and the blood from the superior vena cava passes from the right atrium into the right ventricle and then ultimately to the pulmonary arteries.

The foetal lung is fluid filled and inactive and blood is diverted from the left pulmonary artery through the ductus arteriosus into the descending aorta. At birth when respirations are established the lungs become expanded and an increased amount of blood now reaches the lungs. As the lungs continue to expand with breathing the pulmonary vascular bed enlarges and more blood is then returned from the lungs to the left side of the heart. This causes the pressure in the left atrium to increase and exceed that in the right atrium. As a consequence the septum primum and septum secundum become fused. The septum primum and septum secundum are the two partitions which separate the left atrium from the right atrium. The resulting juxtaposition of the two septa closes the foramen ovale which is a normal defect in the septum primum. Because of the resulting increase in oxygen tension in the blood, the smooth muscle in the ductus arteriosus contracts and causes the lumen of the ductus arteriosus to become obliterated usually within the first two weeks after birth. Once the ductus closure has been achieved there is complete separation of the pulmonary and systemic circulations. The foetal to adult transition has now taken place and the adult circulatory state has been achieved at the expense of the foetal circulatory state.

Transitional heart murmurs

These heart murmurs are fairly common but will seldom be audible after 48 hours. Shortly after birth as the pulmonary blood pressure begins to fall there will be a physiological left to right shunt across the ductus arteriosus if this is still patent and the blood flow across the ductus will allow a murmur to be heard and if transitional it will have disappeared by 48 hours. If the murmur persists after that time and baby is pink and not cyanosed then it will usually be due to a structural lesion such as a ventricular septal defect, a persisting patent ductus arteriosus or rarely, a coarctation of the aorta.

Cyanotic congenital heart disease

These are the so called "blue" babies and they have complex structural heart lesions which result in a mixing of oxygenated and deoxygenated blood from the left side and the right side of the heart. Once diagnosed these babies require urgent investigations and if feasible heart surgery.

Acyanotic congenital heart disease

In these situations the baby is pink and a heart murmur is heard although not all heart murmurs are necessarily present at birth. They may appear after a few weeks when a suitable gradient between the left and right side of the heart has been set up.

Chapter 10 : The Chest

Ventricular septal defect (VSD)

This is the most common of the congenital heart defects. The left and right ventricles of the heart are arranged in parallel and are separated by a partition (septum) which is divided into a lower muscular and an upper membranous portion. For the first few hours after birth the pressures in the right and left ventricles are equal.

Over the next hours, days or weeks the pressure in the right ventricle falls because the pressure in the pulmonary vascular bed falls thus allowing a gradient to be set up between the left and right ventricle. If there is a defect in the septum, blood will then pass through this hole from the left side to the right side of the heart. The turbulence caused by the blood flowing through the ventricular septal defect (VSD) allows a heart murmur to be heard. The extra blood in the right ventricle is then re-circulated through the lungs and baby remains pink.

Most ventricular septal defects are small and do not affect baby in any way. About 75% of small defects will close spontaneously over the next months or years especially if present in the muscular part of the septum. A septal defect is classified as intermediate if it is in the peri-membranous part of the septum. Large defects put strain on the ventricles and may lead to difficulty with feeding because of breathlessness associated with the exercise of feeding and thus there will be poor weight gains. These large defects will usually require medical or surgical management although about 10% of large defects can become smaller and eventually close spontaneously. The heart murmur is short and high pitched if the defect is small or if larger the murmur will then be harsh and last throughout the ventricular contraction (pan systolic). The murmur is heard best down the left sternal edge and towards the apex of the heart and the heart sounds are unremarkable in the early stages. The cardiac impulse is not prominent unless some weeks later if the defect is large and there may then be deformation of the chest wall and a palpable heave and thrill.

If a parent has had a VSD there will be a 1 in 25 chance of baby having the same condition. If a sibling has had a VSD then the chance of a subsequent sibling being affected is about 1 in 33.

Aspirin taken early in pregnancy has been associated with a two-fold increase in the frequency of septal defects.

Atrial septal defect (ASD)

This defect occurs in about 1 in 1000 live births. With an ASD blood flows from the left atrium to the right atrium through the inter-atrial defect. After birth inter- atrial shunting does not occur since pulmonary vascular resistance is still high meaning that the pressures in the left and right atrium are similar. As pulmonary vascular resistance falls, pressure in the right atrium also falls and a gradient is then set up so that blood flows through the septal defect from the left side of the heart to the right. Therefore a greater volume of blood circulates through the defect to the right atrium, the right ventricle and the lungs.

The heart murmur is heard at the base of the heart and is soft and difficult to hear. The murmur is not due to blood flowing through the septal defect since the pressure gradient is small and atrial pressures are low but rather because of increased flow of blood through the pulmonary valve to the lungs. The increased flow of blood through the pulmonary valve delays the closure of the valve and consequently the 2nd heart sound which is due to the closure of the aortic and pulmonary valves becomes widely split.

Spontaneous closure of an ASD is quite common and a small defect may close in the first 18 months. An ASD that is still present at 3 years will probably not close spontaneously and elective closure of the defect is usually undertaken at about 5 years of age

In any event patients who have an ASD repair before 24 years have the same life expectancy as anyone else. A later repair carries a shorter life expectancy. Usually there are no symptoms during childhood and early surgery is reserved for severe symptoms in infancy or in older children where there is a large defect and the possibility of pulmonary artery hypertension. Various ways of closing the septal defect are possible. Surgical closure may mean closure with sutures or depending on the size and shape of the defect a patch. A recent development is the Amplatzer septal occluder which is a double umbrella with a scallop shell configuration. This is inserted via a catheter and covers the septal defect by attaching to the atrial septum.

If a parent or a sibling has had an ASD the new baby has a 1 in 40 chance of being born with the same defect.

Patent Ductus Arteriosus (PDA)

The ductus is a muscular blood vessel linking the left pulmonary artery with the descending aorta. During foetal life it acts as a conduit by diverting blood from the lungs via the left pulmonary artery to the aorta. This limits the amount of blood reaching the fluid filled lungs. After birth the ductus under the influence of increased oxygen tension is obliterated. However the ductus can re-open after birth and this is more likely in babies with a birth weight of less than 1000gms and less likely in babies weighing more than 1500gms. In full term babies the ductus has closed in 50% by 24 hours, in 90% within 48 hours and in all babies by 96 hours. In pre-term babies the ductus may remain open for days or weeks. During foetal life the shunt through the ductus is from right to left in order to divert blood from the lungs. After birth if the ductus remains patent the flow is from left to right because systemic pressures are greater than pulmonary pressures and this means there will be an increased flow of blood to the lungs. When the ductus remains open after birth the pressure in the aorta exceeds that of the left pulmonary artery throughout the cardiac cycle. There is therefore a continuous flow of blood through the ductus to the pulmonary artery and this turbulence produces a continuous murmur ("machinery" murmur) which is heard best at the base of the heart. The peripheral pulses then have a slapping or bounding quality and in babies the pulses on the back of the foot and in the palm of the hand (dorsalis pedis & carpal pulses) become easily palpable when normally they are very difficult to feel.

Most children with a PDA are free of symptoms but premature babies can develop heart failure. A large PDA can lead to progressive pulmonary vascular disease because of increased delivery of blood to the lungs. Ducts of all sizes are closed electively to prevent bacterial endocarditis

The first ligation of a PDA was performed in 1938 but these days a transcatheter approach using an occlusion coil made of stainless steel wire and into which are incorporated Dacron strands to enhance ductal thrombosis and closure is used. The choice then lies between a large surgical scar or no scar.

Coarctation of the Aorta

This is a constriction of varying severity in the aorta usually just below or just above the site of insertion of the ductus arteriosus into the aorta

If the coarctation is above the ductus arteriosus while the ductus remains open as in the foetal circulation or in some cases after birth blood will by-pass the obstruction as it passes through the patent ductus. When the ductus closes the constriction in the aorta (coarctation) becomes apparent and a heart murmur becomes audible. A short systolic murmur will be heard over the chest, the left axilla (armpit) and between the shoulder blades. Sometimes in the newborn no heart murmur can be detected. The peripheral pulses in the legs will be diminished compared with those in the arms (femoral pulses weaker than brachial pulses) and there will be a delayed femoral artery impulse compared with the brachial impulse. The feet will feel cooler than the hands and there will be a slower capillary return in the feet. This coolness and delay in capillary return can be made more obvious by lying baby prone and bringing the back of the baby's hand and the back of the foot together and observing the difference. The blood pressure in the arms will be higher than in the legs.

Coarctation of the aorta is uncommon with an incidence of about 1/2000 - 1 /7000. If a parent or sibling has been affected the likelihood of recurrence for further siblings is 1 in 50.

Chapter 11 : The Limbs

Arms and hands

Palmar creases

There are three main palmar creases.

1. A vertical crease which curves around the base of the thumb (thenar eminence). This is known as the "life line".

2. A proximal horizontal crease extending across the palm and commencing at the radial border between the index finger and thumb. This is called the "head line"

3. A distal horizontal crease starting at the space between the index and middle fingers and extending across the palm. This is the "heart line".

A single palmar crease (simian crease) is an amalgamation of the proximal and distal (head and heart line) palmar creases. This is usually a familial trait and is seen in about 2% of babies and can be present on one or both hands. Occasionally it is associated with a chromosomal abnormality such as Down syndrome but in that case other obvious markers will be present.

A Sydney line is an extension of the proximal crease (head line) right across the palm but preserving the distal crease. It is seen in about 5% of babies and again is familial and can be present on one or both hands. It is of no significance.

The single palmar crease (simian crease) is situated more distally on the palm than the Sydney line.

Fingers and thumbs

An extra finger usually an accessory little finger is occasionally present. This may hang by a thin pedicle in which case it can be ligated and will drop off in the next few days. If the pedicle is thick or if it contains an additional bony ray then formal excision when baby is older is necessary.

The 5th finger may be shortened and incurved (clinodactyly) and this is usually a familial finding. The 5th finger may also be short and tapered with a hypoplastic nail and this finding is associated with certain anticonvulsants.

There may be proximal soft tissue syndactyly (fusion) of the fingers where the fingers are joined at the base but without deformation. This is a familial finding. There may also be distal soft tissue syndactyly of fingers and in this case there is associated deformation of the fingers with attached fibrous bands or amputations. Amnionic bands are responsible for these abnormalities.

A bifid thumb is uncommon.

Since circulation is sluggish in the extremities, the hands will be cool and dusky for a week or so.

Forearms

Rarely there can be an amputation of the mid – forearm. This is the result of constriction bands (Streeters bands) which have encircled the limb cutting off the blood supply to tissues and bone, resulting in a clean amputation. The amputated portion depending on the time of the amputation will be resorbed in the amnionic fluid and therefore not necessarily present at the time of delivery

Chapter 11 : The Limbs

The Legs and Feet

Bow legs

All babies are born with tibial torsion (incurved shin bone) which gives them a bow legged appearance initially. Because of intra-uterine compression forces the foetus is carried in a Buddha-like position with the hips and knees flexed and the tibia and ankles rotated inwardly in relation to the femur. This causes the medial portion of the capsule of the knee to become contracted. However after birth with free movement of the legs and growth the contractures become stretched. This bowing of the legs is due to twisting of the tibia which is really just a packaging defect due to intra-uterine compression and one which will resolve spontaneously.

The Toes

Check that there are five toes on each foot. An accessory little toe may be seen occasionally and if the hands and feet each contain six digits then Ellis van Creveld syndrome is likely, in which case an Atrial septal defect may also be present. Like the thumb the big toe may be bifid. Soft tissue syndactyly of the second and third toes is quite a common familial finding and is of no significance. Frequently especially with the big toe the toenail gives the impression that it will grow through the toe rather than over the top but it never does. A very wide space between the big toe and the second toe can be seen in Down syndrome.

Pes planus (flat feet)

All babies when born have flat feet with no arch and this is because of ligamentous laxity in the supple baby. By 6 years of age 80% of children will have developed an arch. If the child can walk or stand on tip toe the arch will then appear and no treatment will be needed.

Metatarsus varus

The feet when viewed from below have a curved banana-like appearance because the feet are rotated inwardly due to the Buddha-like foetal position. This again is a packaging defect due to intra-uterine compression and gradually resolves over the next weeks or months. The foot can be passively stretched by holding the heel firmly and stretching the foot outwards (laterally) for a few seconds half a dozen times whenever the napkins are changed.

Calcaneo-valgus foot

This is not a club foot and is always positional and always due to intra-uterine compression. The foot is folded back against the outer aspect of the shin and the Achilles tendon is stretched and long. This positional deformation improves spontaneously and fairly rapidly. Passive stretching and over correction of the ankle in order to stretch the dorsal tendons and ligaments can be undertaken by parents if they wish.

Talipes equino-varus

There are two types of talipes equino-varus, the most common being positional talipes due to intra-uterine compression forces and which will require no treatment. The second type is true or structural talipes equino-varus (club foot) which always needs orthopaedic correction or surgery.

Positional talipes

This can always be passively over corrected and there is no foot or calf muscle wasting. Inspection of the calves from behind shows no suggestion of asymmetry. Passive corrective exercises will help bring about spontaneous resolution but nevertheless with time this will occur in any event

Structural talipes ("club foot")

This is the true club foot and cannot be passively corrected. The foot is kidney shaped and the ankle held in an equinus (rotated inwards) position. There will be wasting of the foot and calf muscles and inspection of the calves will show an obvious asymmetry in growth. There is atrophy of the peroneal group of muscles in the anterior muscle compartment. Club foot is due to a combination of intra-uterine compression with inherited genetic factors. It may be unilateral or bilateral.

Structural talipes is twice as common in males as in females and there is an incidence of about 1 in 1000 live births. It is less common in Asians (about half the number) and more common in Polynesians being about 6 times higher. By 9 weeks gestation the foetal foot has skeletal changes consistent with club foot, but by the 11th foetal week the foot is in a normal position.

Structural talipes equino-varus requires treatment from the first week of life, since better correction is possible when the infant's ligaments are lax secondary to maternal oestrogens. A series of 4-8 plaster casts at weekly intervals should provide significant correction. Infants with more severe club foot may need surgical correction with postero-medial tendon release some time later and by 9 months when growth velocity is at its peak. A good outlook for correction can be expected in females with unilateral club foot because female infants have more ligamentous laxity. Joint laxity can be shown by an ability to bend the thumb back on to the forearm, to hyperextend the fingers so that they lie parallel to the forearm and an ability to hyper extend the knees and elbows beyond a straight line. A poor outlook is more likely in males older than 9 months at time of surgery, or with a deep medial crease in the plantar arch of the foot and obvious calf muscle wasting. In adult life the affected foot will remain smaller although shoes of the same size can be worn but the calf will be about 10% thinner. There will be decreased movement and some decrease in strength in the ankle. However children who have had corrected congenital club feet can take part in most if not all athletic activities and with no apparent functional disability despite residual wasting of the foot and leg.

Congenital dislocation of the hips (CDH)

The general incidence of CDH is 2.5-6.5/ 1000 live births compared with in an unscreened population at 1 year of age of 1-1.5/1000. This suggests that in many cases there is spontaneous resolution. There is a family history in about 20% of cases and it is about six times more common in female than male babies.

There are two forms of CDH;

1. Joint laxity - this is recognized at birth.

2. Developmental dysplasia (shallow hip socket) - may be recognized during the first year of life and can lead to eventual dislocation.

At birth a baby's hips may be stable, "clicky", subluxable (unstable), dislocatable or dislocated. The hip is dislocatable when the femoral head is at rest in the hip socket (acetabulum) but can be dislocated by manipulation (Barlow manoeuvre). The hip is dislocated when the femoral head is at rest outside the acetabulum, but can be reduced by manipulation (Ortolani manoeuvre

Developmental dysplasia is diagnosed on ultrasound or X-ray studies where the acetabulum is steeper and shallower than normal

An unstable or subluxable hip is one that is freely mobile and rides up on the acetabular ridge but is not obviously dislocatable – a subjective impression. Clicky hips are due to ligaments riding over bony prominences when they impart a distinct "clicking" sensation and have no bearing on unstable or dislocated hips.

The cause of hip instability or dislocation in the newborn is due to a combination of factors, which encompass genetic, hormonal and positional influences. The frequency of hip instability is about 20% in breech deliveries and 1.5% for vertex deliveries. In the breech position the foetal intra-uterine position is classified as extended when the hips are maximally flexed and the knees extended (50% have hip instability). If the breech foetus has flexed knees and hips then the likelihood of hip instability is about 8%. The left hip is more likely to be dislocated than the right because the foetus usually lies with the left side towards the mother's back regardless of being breech or vertex in presentation. In this position the upper left leg is more constrained by the maternal lumbar spine and therefore more likely to be adducted and as a consequence more easily dislocated.

CDH occurs 5-8 times more frequently in girls than boys. At least 80% of dislocatable hips at birth will resolve spontaneously probably because of adequate post-partum hip abduction with thigh flexion while the ligaments regain their normal tension.

The higher incidence of CDH in female babies is thought to be due to increased ligamentous laxity. There is a preponderance of

Chapter 11 : The Limbs

CDH in first born babies and this is independent of maternal age. CDH is more common in winter months and there is a significant increase in professional and managerial groups.

With normal parents who have had a baby with CDH, the risk for subsequent siblings is about 5% (sisters 10% and brothers 1%). If one parent has had CDH the risk for a child is 12% (sons 6% and daughters 17%). With an affected parent and an affected child the risk for the next baby is 35%.

There are probably two independent gene systems in the inheritance of CDH. One system concerns the development of the acetabulum (hip socket) and is polygenic while the other concerns the laxity of the hip joint capsule as well as the capsules of other joints and is an autosomal dominant characteristic.

In 1937 Ortolani an Italian Paediatrician described the reduction technique where with the pelvis fixed, the combination of thigh abduction and traction coupled with inward pressure over the great trochanter of the femur brought about reduction of the dislocated hip with a palpable "clunk". In 1962 Barlow described a dislocation provocation technique. With the pelvis immobilized and the hip flexed, thigh adduction and downward longitudinal pressure on the femur at the flexed knee joint causes a dislocatable hip to slip posteriorly over the edge of the acetabulum again with a palpable "clunk". The hip is then relocated using the Ortolani technique. A stable hip will not dislocate. "Clicking" hips are due to ligaments sliding over bony prominences and can be felt in about 10% of hip examinations. They do not indicate joint instability and do not need investigations or treatment. Thigh creases may be uneven and one leg can appear shorter due to tilting of the pelvis but provided the hips are stable, then investigations and treatment are unnecessary

Babies who have been breech during the pregnancy, or born breech, or who have a family history of CDH or dysplastic hips or who had a dislocatable hip which became stable in the first few days after birth will all require an ultrasound study of the hips at about 6 weeks of age to exclude acetabular dysplasia. Note that 3- 4% of adults in their mid fifties have osteoarthritis of the hip or hips and about half of them will have had developmental hip dysplasia as a cause.

Ultrasound screening of the hips was introduced in the 1980"s. It is non-invasive and visualizes soft tissue structures such as cartilage which would not be seen with conventional X-rays. It can also be used to observe the stability of the hip during stress manoeuvres. Ultrasound screening of the hips is reliable up until about 1 year of age.

About 5% of babies will have dislocatable hips at birth or dysplastic hip changes on ultrasound but in this group of 5% only about 10% will remain abnormal and eventually need treatment.

The Pavlik harness which is a malleable Aluminium frame has become the standard treatment for CDH in infants under six months of age and who have a clinically reducible hip. The success of the harness lies in its ability to maintain hip flexion at about 70 degrees of hip abduction while allowing some free rotation of the hip to occur. The harness is adjusted so that the hip does not extend beyond 90 degrees and it does not force abduction but allows progressive stretching of the hip adductors by the weight of baby's leg. The harness allows active hip movement but prevents the infant from fully extending the hip in an adducted position. An X-ray of the pelvis with the harness in place is needed to ensure that the hip has been reduced and in place and that the flexion of the hip greater than 90 degrees has been achieved. The harness is removed at two weekly intervals to allow for bathing of the baby once the hip has become stable. Treatment in the Pavlik harness is usually required for three months. The Pavlik harness was developed by Arnold Pavlik during the Nazi occupation of Czechoslovakia during the 2nd world war. It was seen as a new functional method in which the most important stage of the treatment was done by the baby within the harness. The principle was based on hip and knee flexion producing hip abduction and then gentle reduction. The use of the harness lead to a decrease in avascular necrosis of the femoral head and a decrease in the incidence of femoral nerve palsy both of which were due to hyperflexion of the hip.

Boys with CDH have a higher rate of treatment complications than girls. There is a higher risk of re-dislocation and while 90% of girls have success with the harness only about 10% of boys have satisfactory reduction. Closed or open reduction should be used if a hip is not reduced after four weeks full time in the Pavlik harness and again males have a less likely success rate after a closed reduction than females.

The use of double or triple napkins is not adequate treatment for CDH since this cannot ensure hip flexion which is one of the important components in proper positioning of the femur. Double napkins can be used as an interim measure while waiting for application of the Pavlik harness. An irreducible hip and an age greater than six months are usually contra-indications to the use of the Pavlik harness. These cases require closed reduction under general anaesthesia and plaster cast immobilization.

Chapter 12 : The Genitalia

Males

The Penis

The body of the penis is composed of three elongated masses of erectile tissue (right and left corpora cavernosa and the corpus spongiosum). The glans penis is attached to the corpora cavernosa and the urethra which lies within the glans opens as a sagittal slit near the tip of the apex of the glans. At the neck of the penis the skin which is very thin and loose is folded upon itself to form the prepuce (foreskin) and this overlaps the glans for a variable distance. Mild torsion of the penis may be seen in 1-2% of newborn infants, and this is a normal variant.

The external meatus may be located just below the tip of the glans penis for some male infants but for most it is at the tip.

Penile variants

The normal penile length in the newborn is 3.5cms+/- 0.7cms. There may be several reasons for an inconspicuous penis.

1. Buried penis due to an over abundance of pre-pubic fat.
2. Webbed penis due to a midline skin web at the angle between the penis and the scrotum.
3. Trapped penis which is held back by a Dartos band
4. Poorly suspended penis.

5. Diminutive penis which may be malformed because of epispadias, severe hypospadias or because of a chromosomal abnormality.

6. A small penis which is not malformed and which therefore may be related to an endocrine abnormality.

Circumcision

There is a slow and natural separation of the foreskin from the glans penis. In the newborn the inner layer of the foreskin is adherent to the glans so that in over 95% of newborn male babies the foreskin cannot be retracted. By 6 months of age the foreskin can be retracted in 15% and by 3 years of age only 10% of boys will still have an unretractable foreskin

The epithelium which covers the glans and the inner layer of the foreskin gradually matures and separates. The epithelial cells which are shed produce a whitish material called smegma which naturally migrates to the tip of the foreskin. The foreskin should not be forcibly retracted because this will lead to tearing and bleeding, secondary infection, scar formation and later phimosis when the foreskin cannot then be retracted.

It is said that circumcision is an operation performed on a non-consenting patient at the instigation of a medically unqualified 3rd party. The prevalence of circumcision varies widely from one country to another. Prior to 1900 most males were uncircumcised.

The shift towards circumcision was given impetus by a British surgeon James Hutchinson who wrote a paper in 1891 entitled "on circumcision as a preventative of masturbation"

Today circumcision is mainly an emotional issue with the usual reasons advanced in favour of circumcision being custom, religious preference, family patterns, prevention of infection and finally physician recommendation on the basis of phimosis, paraphimosis or recurrent balanitis. In the newborn however circumcision can be seen as cosmetic surgery. After the procedure most infants void within 5-6 hours and 95% will have passed urine by 10-11 hours.

Common complications of circumcision include bleeding, infection and meatal stenosis and the complication rate ranges from

0.2% - 5%. Circumcision appears to be beneficial to those living in tropical countries and in desert areas or where good hygiene is difficult to achieve. However in the past few years as the circumcision rate has decreased evidence is beginning to mount suggesting that there are some benefits for the procedure after all. Urinary tract infections are more common in uncircumcised males, and up to 80% of urinary tract infections in boys under 5 years occur in the uncircumcised. Circumcision may prevent the foreskin being colonized with uropathic bacteria in infancy and childhood. Pyelonephritogenic fimbriated E.coli bind avidly to the inner lining of the foreskin within the first few days of life. Bacterial adherence and colonization of the foreskin is followed by peri-urethral colonization leading to ascending urinary tract infections. The incidence of urinary tract infections is about 0.5% in females, 0.2% in circumcised males and 4% in the uncircumcised; ie; about twenty times more frequent in the uncircumcised than the circumcised male. The bacteria most commonly isolated from the foreskin of adult males are group B streptococci (the usual cause of severe neonatal sepsis) which can cause balanitis and also be transmitted sexually. Uncircumcised men appear to be more susceptible to sexually transmitted diseases that disrupt epithelial surfaces such as genital herpes and condyloma acuminatum (human papilloma virus).

Penile cancer is almost unheard of in males who have been circumcised at birth. The predicted life-time risk of penile cancer developing in an uncircumcised male is about 1 in 600. Uncircumcised men have a 5-8 times greater risk of becoming infected with AIDS virus during unprotected sexual intercourse with an affected woman than do circumcised men.

Epididymitis is frequently associated with urinary tract infection and almost 3/4 of cases of epididymitis are associated with the uncircumcised state.

The infant circumcised without anaesthesia feels pain evidenced by crying, irritability and a varied sleeping pattern. There will be an increase in heart rate and blood pressure but these changes are short lived lasting only minutes to hours. There are no long term psychological effects. Sucrose can relieve pain by inducing the release of endorphins (nature's opiate) from the brain. The Jewish Mohel used to chew dates and spit the saliva into baby's mouth or put a finger previously dipped in sweet wine on the baby's tongue immediately before the procedure. A sucrose solution can be given to baby a few minutes before circumcision.

In the 1980's 70-90% of boys in the USA were circumcised while the rate in Great Britain was 6.5%. In Scandinavian countries the circumcision rate is about 2%. Only about 10% of the world's male population are circumcised.

In the past few years the anti-circumcision movement has grown with a consequent reduction in circumcisions. At present the benefits of circumcision outweigh the risks of the procedure, but most decisions will be based on parental preference or religious conviction. There is also the feeling that what nature has provided should be left alone.

Hypospadias

In this condition the external meatus may lie anywhere from just below the tip of the penis, to anywhere on the shaft, to the perineum. The more proximal (nearer the perineum) the external meatus, then the more likely there will be an associated ventral curvature of the penis because of the presence of chordee, a tight band of fibrous tissue which replaces the normal Dartos fascia. Hypospadias occurs because the development of the external genitalia is arrested before full differentiation, resulting in incomplete formation of the anterior urethra.

In most cases the hypospadias lies on the glandular or coronal part of the penis with a very small minority of cases being somewhere on the penile shaft. Hypospadias is always associated with a partial natural circumcision. Undescended testes and inguinal hernias are the most commonly associated anomalies while a bifid scrotum or a small penis are rarely found. The more severe hypospadias is more likely to be associated with other external genitalia malformation. Since the external genitalia are formed later in gestation than the upper urinary tract the association between hypospadias and upper urinary tract anomalies is infrequent.

If the testes are impalpable in a term baby with hypospadias, then chromosome karyotyping is warranted while a micropenis suggests that pituitary evaluation should also be considered. If a baby with hypospadias has bilaterally descended testes in a well differentiated scrotum the chance of finding any underlying or associated anomalies is slender.

Other possible associations with hypospadias are seen in small for gestational age babies, advanced maternal age and in infants born to infertile parents who conceive by in vitro fertilization.

Since the 1970's there has been a 2-5 fold increase in the prevalence of hypospadias which has largely been unexplained. It has been suggested that foetal exposure to progestins that have been taken as contraceptives by mother, or as part of a pregnancy test or for the maintenance of pregnancy may be responsible. However current theories also suggest that widespread pollution of the environment by synthetic compounds with oestrogenic or anti-oestrogenic activity is rather more likely.

Repair of hypospadias is usually carried out after 12 months of age but this age is lessening and some one stage repairs are undertaken between 3-6 months. The foreskin is used in the construction of the urethra and it is important that these boys do not have a further circumcision before the procedure.

The testes

The testes may be palpable and fully descended or they may be undescended (cryptorchidism) where they are then either palpable but partially descended or impalpable.

By the 15th week of gestation the development of the external genitalia is complete and the foetal testes lie in the intra-abdominal position. Between the 7th- 9th foetal month the testes begin to descend into the scrotum through a canal (processus vaginalis) which later is obliterated. If the processus remains patent then later an inguinal hernia or a hydrocoele will develop. The impalpable testis may be truly impalpable or be retractile and thus remain in the scrotum intermittently. Retractile testes will have a normally developed scrotum while the scrotum will be very small in the truly impalpable testes situation. The undescended testis may reside within the intra-canalicular canal (processus) and be palpable, or remain within the abdomen and be impalpable. It may also be ectopic and lie beyond the external inguinal ring but be palpable and reside within the

Chapter 12 : The Genitalia

superficial inguinal pouch. If both testes are impalpable with a poorly developed scrotum in a full term baby then ultrasound studies are indicated to exclude anorchia (absent testes).

Undescended testes are quite common occurring in about 3% of full term male babies and in up to 30% of premature boys. In about 10% of cases the disorder is bilateral. By one year of age 1% of boys will still have an undescended testis or testes. Most spontaneous descents have occurred within three months of the expected date of delivery of the baby but spontaneous descent is unlikely after six months. Undescended testes are associated with a higher incidence of testicular torsion, inguinal hernias and testicular malignancy. In the older child the best way to check for descent or not is for the child to sit with the hips and knees flexed. This position will prevent an active cremasteric reflex (stroking inner border of thigh) from pulling the testes out of the scrotum.

Since the 1970's the operation to fix the testes in the scrotum has gone from the 5th year of life to the 2nd. The number of spermatogonia in the undescended testis rapidly decrease in the 3rd year and there is lack of growth in the seminiferous tubules. By five years the volume density of the spermatogonia has decreased by about 60%. Fertility is only likely if the sperm count exceeds 60 million/ml. Between 1938 and 1990 the sperm count in males has apparently fallen from 113 million/ml to 66 million/ml. The purpose of surgery for undescended testes is to reduce the risk of malignancy which may be 20-40 times greater than in the normal population and also to ensure fertility. In about 10% of males with an undescended testis a tumour is also possible in the normally descended testis.

If a testis can freely enter the scrotum surgery is not indicated. If a testis can be brought into the scrotum without tension and without fixation the term used is orchiochalasia but if the testis has to be anchored in the scrotum the procedure is called an orchidopexy. However bringing an undescended testis into the scrotum does not necessarily alter the natural course of development of malignancy since most of the late descended testes are dysplastic and probably infertile.

Torsion of the testes

Testicular torsion occurs when the testis lacks its usual attachment to the posterior wall of the scrotum and twists one or more times on the spermatic cord. Undescended testes are more prone to torsion. With testicular torsion there is initial pain due to venous distension followed later by testicular pain due to deprivation of the blood supply. The scrotum will become darkly coloured, tense and the testis will be elevated. The cremasteric reflex will be absent and this is the most reliable clinical sign. If the testis dies it will become hard and painless and the scrotum will be dark blue. If there is any chance of salvaging the testis immediate surgery is indicated. While waiting for surgery an icepack should be applied and an attempt made to unravel the spermatic cord and relieve pain and improve blood supply. To do this the testis is rotated laterally (outwards). The salvage rate falls from 80% if surgery is undertaken within 5 hours of onset of pain to 20% within 10 hours. After 24 hours all testes that have undergone torsion will be non-viable. Generally if a male baby is born with a hard enlarged discoloured and painless testis then it is already too late for surgery to restore viability.

Hydrocoeles

A hydrocoele is a collection of serous fluid in the sac of the tunica vaginalis which surrounds the testis. This is the very common vaginal hydrocoele seen in most newborn males. The tunica vaginalis is the lower portion of the processus vaginalis which precedes the descent of the testis from the peritoneal cavity into the scrotum. Hydrocoeles may be quite large giving a spurious impression of scrotal or testicular size and all hydrocoeles transilluminate brilliantly.

There are several other types of hydrocoele. A hydrocoele may communicate with the peritoneal cavity through a patent processus vaginalis, it may be confined in the processus because the processus is closed at the upper end (deep inguinal ring) or the hydrocoele may be encysted in the processus which is closed at the top and bottom ends. The processus vaginalis will close spontaneously in time and the hydrocoeles then disappear although an encysted hydrocoele may persist. If an encysted hydrocoele is still present after a year or two it could then be repaired surgically.

Females

In a full term female the labia majora will cover the labia minora completely but the earlier the gestation the more widely separated the labia majora appear and the labia minora are prominent and protrude. A mucoid discharge may issue from the vulva and this due to an oestrogenic effect from mother. Sometimes a little blood may normally be seen in the discharge. Maternal oestrogens will make the labia majora appear swollen.

Fusion of Labia minora (synechiae vulvae)

There may be complete or partial fusion of the labia minora in the mid-line by a translucent membrane. This membrane will appear thicker posteriorly. Synechiae vulvae are usually seen in the first 2 years of life and can be present at birth. Generally there are no symptoms although sometimes there can be difficulty with micturition. Gentle lateral manual traction on the labia minora will rupture the membrane and the application of Dienoestrol cream will prevent recurrent fusion and can also be useful in initial treatment.

Hymen

All female infants are born with hymenal tissue and about 1 in 1000 will have an imperforate hymen. A small tag is often seen protruding posteriorly between the labia minora and this is a hymenal tag which will later atrophy. In a full term baby the vagina is about 4cms long.

A cystic swelling of variable size is occasionally seen between the labia minora. This swelling is due to the accumulation of cervical secretions which distend the vagina and cause an imperforate hymen to bulge. The contents of the "cyst" are a combination of cervical mucoid secretions with or without blood. They fall into one of four categories. Maternal oestrogens cause an excessive intra-uterine stimulation of the infant's cervical mucous glands.

1. Hydrocolpos - fluid distension limited to the vagina.

2. Haematocolpos – an accumulation of blood limited to the vagina.

3. Hydrometrocolpos – distension of the vagina and uterus by secretions other than blood.
4. Hydrohaematometrocolpos – distension due to the accumulation of fluid and blood in the uterus and vagina.

In these cases spontaneous rupture may occur in the first few months but if not then surgical opening of the "cyst" should be undertaken.

Ambiguous genitalia

In the female foetus abnormal androgen exposure leads to varying degrees of clitoral enlargement and fusion of the labia.

(a) Congenital adrenal hyperplasia

The female baby will appear virilised with a prominent clitoris and fusion of the labia resembling the scrotum but with no palpable gonads. There will be one of several possible enzyme deficiencies leading to an accumulation of steroids proximal to the enzyme block and the consequent conversion of these steroids to androgens with a resulting virilising effect.

(b) Intersex anomalies

These are rare and caution should be exercised before giving baby a name and a definitive course of action has been decided upon. There will be virilization of the external genitalia but in these cases gonads (ovo-testes) will be palpable in the fused labio-scrotal folds.

Chapter 13 : The Back, Abdomen and Anus

The back

Coccygeal dimples (pits)

These are common and are seen in about 5% of babies. They are situated in the gluteal (sacral) cleft and are orientated downwards (caudally). They are not associated with any intradural pathology. As a consequence they do not need investigating. Quite often there is a familial pattern.

Sacral dimples (pits)

These are very uncommon and are situated above the gluteal cleft. The dimple is orientated upwards (rostrally) and if deep enough can be associated with intradural pathology such as a tethered spinal cord. If a true sinus is present it can be situated anywhere along the spine from the occiput to the sacrum but most are in the lumbo-sacral area.

Sacral dimples can be classified in several ways.

- a dimple - a shallow depression above the gluteal crease and with an easily visible base.
- a dimple in the gluteal crease with clearly visible intact skin at the base.
- a deep dimple requiring skin retraction and magnification to see the base.
- a presumed sinus which is a deep dimple with the base not able to be seen.

- a mass – a mound of skin surrounding a deep dimple with an invisible base.
- a mid-line discoloured pigmented area with a café au lait patch or an haemangioma.
- a skin tag in the midline
- a midline hair tuft in association with a deep dimple.

Dimples, above or in the gluteal crease or deep dimples with a clearly visible base do not need investigating unless there is an associated skin mass, skin tag, pigmentation or abnormal tuft of hair. A deep dimple with the base not able to be visualized also requires investigation. Initial investigations are usually ultrasound, X-rays and probably MRI.of the lumbo-sacral spine.

Dermal sinuses are epithelial lined tracts that terminate at a deeper level and can communicate with the spinal cord (tethered cord). They arise because the epithelial ectoderm and the neuro-ectodermal epithelium fail to separate when the neural groove closes to form the medullary tube. At birth the spinal cord conus is situated at a lower level than that of an adult but achieves the adult level by the end of the first year of life unless the cord is tethered (spinal dysraphism). After one year the difference in the growth rates between the vertebral column and the spinal cord approaches zero. Thereafter the relative position between the spinal cord and the vertebral column at any given segment remains constant during subsequent lengthening of the trunk.

The abdomen

Umbilical hernia

An umbilical hernia occurs when there is a failure of the normal dense fibrous connective tissue to occlude the umbilical ring after the birth. The peritoneum protrudes through the resulting defect but it is covered with skin. The protrusion can easily be reduced and the umbilical ring will then be palpable.

Umbilical hernias are seen in about 4% of babies under 6 weeks of age and can vary in size from 1cm upwards. They are twice as common in females and more likely in babies born with a thick umbilical cord or a premature baby weighing less than 1500gms.

Most umbilical hernias are not associated with any symptoms and will close spontaneously and do not need to be taped. However if the hernia is showing no sign of closure by 4 years of age then surgical closure may become necessary since spontaneous closure is unlikely after 4 years.

Supra-umbilical hernia

This is seen as a small defect in the linea alba in the mid-line above the umbilicus. It is of no consequence and eventually will close and disappear.

Inguinal hernia

Oblique inguinal hernia:

The processus vaginalis is a peritoneal pouch which precedes the descent of the testis from the abdomen into the scrotum. There may be a congenital defect in the processus which normally is closed at birth or closure may be completed later. Closure occurs first at the deep inguinal ring and at the top of the epidydymis and gradually extends until the processus is turned into a fibrous cord. If the processus vaginalis remains patent the herniated gut will descend in front of the testis into the tunica vaginalis. The processus vaginalis and the tunica vaginalis constitute the sac of the hernia.

Inguinal hernias are rare at birth and are seen in about 1.5/100 live births. They are less common in breast fed babies. They are characterized by failure of the processus vaginalis to fuse during the last few weeks of gestation and hence are more common in premature babies. Unless the hernia is symptomatic by being difficult to reduce or by becoming strangulated, surgical repair is best left until the baby is about 46 weeks post conception age. Earlier repair is associated with an increased risk of apnoea during the first 12-18 hours post anaesthesia. Also the hernias are difficult to repair because the hernial sac is thin.

Direct inguinal hernia

These are uncommon and are always acquired and are more common in men.

Femoral hernia

This is very rare in the newborn and is more likely in females. A defect in the femoral ring allows the intestines and sometimes an ovary to protrude through the abdominal wall.

The Anus

The anus is centrally placed in the pigmented skin of the perineum and its appearance hardly changes between birth and adulthood. After puberty hairs develop in the pigmented skin of the perineum in males.

Anterior anus

A slight anterior placement of the anus is not uncommon. The anus will appear to be placed eccentrically in the pigmented anal skin. No treatment is required unless constipation which is a possibility becomes a problem.

Anal membrane

In a normally placed anus a green bulging mass due to accumulated meconium will be seen behind an overlying anal membrane. A simple incision is usually sufficient to allow the passage of meconium.

Imperforate anus

This should be looked for immediately after the baby is born. It will appear as a dimple in the skin where the anus would usually be found and of course no meconium can be passed. Relevant investigations will determine the degree of severity of an imperforate anus.

Perineal membrane

In some female infants a translucent membrane will be seen lying in the mid-line between the fourchette and the anus. This membrane will slowly become epithelialised and no treatment is necessary.

Chapter 14 : Jaundice

Most newborn babies will become jaundiced (yellow discolouration of the skin and the eye balls). Jaundice is due to a pigment bilirubin that has been produced from haemoglobin which is contained in red blood cells which after birth have become redundant. These redundant red cells break down normally or else are haemolysed which means they are destroyed more rapidly than normal. The breakdown of 1gm of haemoglobin will yield 34mgms of bilirubin and the destruction of red blood cells accounts for 75% of the daily bilirubin production in the normal newborn baby. The normal newborn baby will produce 8.5mgms of bilirubin/kgm body weight / day. Red blood cells in the newborn have a life span 2/3 that of an adult unless they are being destroyed rapidly which is usually by haemolysis.

When haemoglobin is broken down it is the heme component which releases biliverdin initially and which is then rapidly converted into bilirubin. In the circulation bilirubin is bound avidly to a protein albumin which transports the bilirubin to the liver. The very small amount of unbound bilirubin left is insoluble in water but readily fat soluble and it is this unbound fraction which can penetrate the blood-brain barrier. Bilirubin bound to albumin is in an unconjugated form. When bilirubin has been processed in the liver it becomes conjugated and is now water soluble and is transferred to the bowel in the bile. In the small bowel an enzyme (bilirubin ß-glucuronidase) may deconjugate the bilirubin thus allowing it to be re-absorbed back into the blood. This is called the entero-hepatic circulation. Bilirubin that was not de-conjugated in the bowel passes on and is responsible for the yellow colour of the bowel motions.

Chapter 14 : Jaundice

Transient unconjugated hyperbilirubinaemia (jaundice) occurs in all newborn babies in the first week of life and for several reasons. The red cells have a shorter life span, there is a higher circulating red blood cell volume and a greater entero-hepatic uptake with re-cycling of bilirubin. The liver system is immature in terms of bilirubin uptake, conjugation by enzymes in the liver and biliary excretion. In the foetal circulation, the placenta is unable to clear conjugated bilirubin but is far more permeable to unconjugated bilirubin and thus help in the clearance of heme breakdown products.

Physiological jaundice is beneficial for the baby. The bilirubin pigment induces liver enzymes making them more efficient and bilirubin itself is a powerful anti-oxidant, cleaning up free oxygen radicals and helping to protect against infection. Therefore there is no need for baby to be placed in the sun in an effort to lessen physiological jaundice.

However in certain clinical situations bilirubin can predispose to brain toxicity (bilirubin encephalopathy). If the concentration of bilirubin is high enough the binding capacity of the transport carrier albumin will be exceeded and free bilirubin can then pass across the blood-brain barrier. The blood-brain barrier can also be disrupted by certain events which then allow bilirubin-albumin complexes to enter the brain. In the newborn baby these events are most likely to be due to hypoxia (insufficient oxygen), hypercarbia (excessive carbon dioxide), acidosis and hyperosmolar conditions such as are encountered in a sick premature infant.

The auditory pathway is vulnerable to damage from moderate to severe jaundice resulting in a sensori-neural hearing loss. A birth weight <1500gms and prolonged jaundice and acidosis are precursors for a sensori-neural hearing loss. However in healthy full term jaundiced infants without evidence of haemolysis there have been no consistent effects upon intelligence, neurological sequelae or hearing.

As a rough estimate the bilirubin level in the newborn baby may be estimated as follows.

- - jaundice from head to collar bones - 85si units
- - jaundice down to umbilicus - 130si units
- - jaundice down to knees - 200si units
- - jaundice down to ankles - 250si units

- jaundice of palms and soles of feet - >250si units

Bilirubin toxicity is rare in infants without haemolysis and it has been suggested that treatment (phototherapy) in these cases be deferred until bilirubin levels are between 400-500si units. If however there is evidence of rapid red cell breakdown (haemolysis) then bilirubin levels should be kept below 300-400si units. At present a rather more aggressive approach has been taken and treatment regardless of haemolysis or not is initiated at the following values.

- first 24 hours -135si units,
- 24-48 hours - 225si units
- 48-72 hours – 250si units
- 72-96 hours – 275si units
- 96 hours or more – 300si units.

In full term babies bilirubin levels peak between 3-5 days and jaundice is not detected in the first 24 hours unless there is evidence of haemolysis when there is an accelerated breakdown in red blood cells. The onset of jaundice after 24 hours is usually physiological. It should be remembered that bilirubin even when bound to albumin is a very powerful endogenous anti-oxidant comparable with such exogenous anti-oxidants as vitamin C and vitamin E. Breast feeding by helping to promote jaundice will compensate for any possible anti-oxidant deficiency and offer protection for baby.

If the arbitrary treatment levels for jaundice mentioned earlier are exceeded then treatment in the form of special wave length light (phototherapy) is implemented. Exchange transfusions of blood which were common in the days before phototherapy are now uncommon and are reserved for severe jaundice unlikely or failing to respond to phototherapy. As a yellow pigment bilirubin can only absorb light in the blue, green and violet range. Under laboratory conditions blue light (wave length 450nm) is best absorbed and is twice as effective as green light (wave length 500nm) although this is not necessarily the case in clinical conditions where green light is more likely to pass through the skin unabsorbed and unscattered compared with blue light. With phototherapy bilirubin is converted mainly into a product called lumirubin which is harmless and being

Chapter 14 : Jaundice

water soluble can by-pass the liver and be excreted directly into the urine and bile. During phototherapy a mixture of bilirubin and lumirubin is being measured so that there will not be an obvious and rapid decrease in the bilirubin level.

Treatment consists of placing baby in a fibre-optic suit or on a fibre-optic blanket which is probably no more effective than being nursed under an overhead light source (phototherapy unit). However with fibre-optic phototherapy there is no need for eye cover and baby remains in the cot by mother's bed with better mother-baby contact. Sometimes double lights, a combination of a fibre-optic blanket and an overhead light unit are used.

There are some minor side effects with phototherapy such as an increase in insensible water losses, diarrhoea with green stools or a skin rash all of which are reversed by discontinuing phototherapy. Pigmentation of the primary dentition may result from the deposition of bilirubin into enamel and dentine. Bilirubin later becomes oxidized to green biliverdin leading to a yellow, green or blue discolouration of the teeth. The intensity of the colour is directly proportional to the severity and duration of the jaundice. The discolouration fades with time and the affected teeth gradually revert to their normal colour.

Breast milk jaundice

Certain criteria must be met for jaundice to be called breast milk jaundice.

- there must be a rising bilirubin level beyond the 4th day of life in the absence of a haemolytic disorder, or there must be prolonged jaundice which can persist for up to 3 months in a thriving solely breast fed baby.
- if breast feeding is suspended for 24 hours there must be a decline in bilirubin levels of 80-170si units in that time.
- there should be a rise in bilirubin levels but to a lesser degree when breast feeding is resumed.

Breast feeding should not be discontinued because of breast milk jaundice unless bilirubin levels are very high (over 400si units). If breast feeding is to be suspended it should not be for more than 24-48 hours. The incidence of breast milk jaundice is about 1 in 50 for solely breast fed babies and there may be a 70% recurrence rate in subsequent siblings.

Breast fed babies have higher bilirubin levels than formula fed babies. Breast feeding will increase gut motility and decrease the rate of intestinal re-absorption of bilirubin. But if breast milk is rich in an enzyme beta glucuronidase then the conjugated bilirubin in the gut will be deconjugated by this enzyme and in this form it is readily and easily re-absorbed back into the circulation (entero-hepatic circulation). This follows then that bilirubin levels tend to increase and the re-cycling process will cause prolonged jaundice. Remember that breast milk jaundice is beneficial and harmless rather than harmful because bilirubin is inducing liver enzymes and acting as a powerful anti-oxidant. There is no need to place baby in the sun in an attempt to lower bilirubin levels.

Jaundice due to Haemolysis

ABO and Rhesus incompatibility

The red blood cell survival time in newborn full term babies is between 70-90 days and in premature infants 50-80 days. Red blood cells in children and adults survive for 100-120 days. Red blood cells are able to deform their outer membrane and their intracellular contents in order to be able to pass through very small capillaries (micro-circulation). Thus if there is any abnormality in the red cell shape, its metabolism or haemoglobin structure there will be decreased red cell membrane flexibility and these cells will be removed and destroyed in the spleen and liver, their destruction giving rise to an increase in bilirubin production.

ABO Haemolytic disease

This haemolytic condition was first described in 1944. In 15% of all pregnancies in Caucasian women an ABO situation will arise with mother being group O and baby either group A or B. In 60% there will be OA incompatibility and in the remaining 40% OB incompatibility. However in only 10-20% of ABO incompatibility infants will there be significant jaundice. The Coombs test which denotes red cell sensitization will be positive in 25% of ABO incompatibility babies but will give no indication of the degree of jaundice. Usually there is no suggestion of anaemia but the blood film will show an increase in red cell production (reticulocytosis) to compensate for increased destruction and small round red cells (micro-spherocytes) will also be seen.

Group O mothers will naturally form antibodies against group A or B red cells and these antibodies are of the IgG class and are therefore able to cross the placenta into the foetal circulation. These antibodies have a multitude of A and B binding sites to attach to in the baby but only a small number of these binding sites are on the foetal red cell membrane. The small number of binding sites on the red cell membrane means that the Coombs test may be negative or only weakly positive.

IgG anti A and anti B antibodies are divided into three sub groups namely IgG1, IgG2 and IgG3 the most common being IgG2. However IgG2 after crossing the placenta does not bind onto the white cell receptor sites which would carry it to the antigenic sites on the red cell membrane and therefore this sub group antibody is incapable of destroying red blood cells. On the other hand IgG3 and to a lesser extent IgG1 antibodies which are present in low concentrations in foetal blood do bind to white cell receptor sites and have strong red cell destructive activity with IgG3 being the more effective in lysing red blood cells. This fact would seem to account for the cases where the Coombs test is negative or only weakly positive and yet there is significant jaundice due to haemolysis. This is the main reason why serological tests are unreliable in predicting the degree of jaundice in the baby. Of the three classes of IgG antibodies the sub class with the highest titre which is usually IgG2 will determine the result of the Coombs test and IgG2 is incapable of causing red cell lysis. In addition IgG1 and IgG3 can vary greatly in their red cell lysing ability in the serum of different mothers.

In group A and group B mothers the naturally formed antibodies anti B and anti A respectively are of the IgM class and unlike IgG antibodies, IgM antibodies cannot cross the placenta, and therefore are unable to reach the foetal circulation and cause haemolysis with red cell destruction. Jaundice in these babies is not due to haemolytic disease.

Rhesus haemolytic disease

This uncommon haemolytic disease occurs when a Rhesus positive baby is born to a Rhesus negative mother. There are several Rhesus (Rh) antigens (Cc,D,Ee) each of which can be detected by a specific antibody. The most important of these antigens is the D antigen and red blood cells which possess the D antigen are Rh positive. The symbol "d" denotes the absence of "D" and means that the red blood cells are Rh negative. No anti-d antibody has been found and there is thus no specific "d" antigen. Individuals who have the genotype "DD" or "Dd" are therefore Rh positive while those who are "dd" are Rh negative. Hence the various blood groups are designated O+ve or O–ve, A+ve or A-ve, B+ve or B–ve and AB+ve or AB-ve.

About 15% of the Caucasian population are Rh-ve while the Rh-ve genotype is virtually non-existent in those of Oriental ethnicity.

Rh haemolytic disease occurs in about 1% of first pregnancies involving a Rh negative mother and a Rh positive foetus. The likelihood of the Rh negative mother having an affected baby increases with each pregnancy in which the foetus is Rh positive. Small volumes of foetal blood enter the maternal circulation throughout the pregnancy and especially in the last trimester. But the major amount of foeto-maternal transfer of blood and which is responsible for most of the sensitization occurs during the delivery. The initial response of the Rh negative mother to the infusion of foetal Rh positive red blood cells is to produce IgM anti- D antibodies and of course IgM antibodies cannot cross the placenta. However this response is later followed by the production of IgG anti-D antibodies and being IgG antibodies they can cross the placenta and destroy the foetal Rh positive red blood cells, the amount of destruction due to this haemolytic process depending on the amount of anti D antibody that has been transmitted. The anti D antibodies that mother has formed actively in this response are permanent and do not decay unlike passively acquired antibodies. They remain to affect the next Rh positive baby.

Chapter 14 : Jaundice

The requirements for Rh haemolytic disease can be summarized as follows:

(a) Rh negative mother and Rh positive foetus (baby).

(b) leakage of foetal red blood cells into maternal circulation.

(c) maternal sensitization to the D antigen on the foetal red cells.

(d) results in production of maternal anti D antibodies.

(e) transplacental passage of maternal anti D antibodies into foetal circulation.

(f) attachment of maternal anti D antibodies to Rh positive foetal red blood cells.

(g) destruction of the anti D antibody coated foetal red blood cells.

(h) jaundice due to release of bilirubin from destroyed foetal red cells.

It had been found that the frequency of Rh haemolytic disease was much lower in ABO incompatible pregnancies. The naturally occurring anti A or anti B antibodies in mother's serum caused the destruction of placentally transmitted foetal A or B red blood cells before sensitization could occur. On the basis of this finding attempts were made to prevent maternal Rh sensitization by giving mother an injection of potent anti D serum which would then destroy transmitted foetal Rh positive red blood cells before there was time for sensitization to occur.

This procedure has proved so successful that all Rh negative mothers provided that they have not already been sensitized in the past are routinely given anti D serum (Rhogam) after the delivery of an Rh positive baby. The use of the commercially available product beginning in 1969 has achieved the expected results so much so that that exchange transfusions for Rh incompatibility which were very common are now rare. Anti D immunoglobulin is manufactured from anti D antibodies made in Rh negative male donors who have been sensitized by repeated transfusions of Rh positive red blood cells. The 300mcg of anti D in a standard vial should be sufficient to render 15mls of Rh positive red blood cells (30mls of blood) non-immunogenic. This single intramuscular injection should be given within 72 hours of the delivery of an Rh positive baby. The passively acquired anti D antibody slowly disappears although it can be detected in maternal serum six months later. If anti D (Rhogam) is not given post partum the maternal sensitization rate is about 15% but if it is given then this rate falls to about 1.5-2%. Some of the failures despite giving anti D immunoglobulin (Rhogam) to the mother are due to a foeto-maternal bleed of more than 30mls of blood. By testing mother's blood after the delivery the size of the foeto-maternal transfusion can be estimated and the dose of Rhogam adjusted accordingly. The administration of Rhogam will do nothing to prevent sensitization from previous transfusions, miscarriages or missed Rh positive pregnancies, since the actively produced anti D by mother is permanent. In other failures there may have been foeto-maternal during the pregnancy and usually in the third trimester thus allowing mother to make natural anti D antibodies before the delivery.

Another interesting possibility for failure of Rhogam is the grandmother theory. If the maternal grandmother is Rh positive, she may sensitize her Rh negative daughter at birth when her Rh positive red blood cells enter the baby's circulation. The new Rh negative female infant will then make anti D antibodies in the first months of life. These antibodies will be permanent of course and this infant when she grows up will be found to have been apparently inexplicably sensitized and therefore sensitize her new (3rd generation) Rh positive baby.

The haemolytic process results in the production of bilirubin and the baby will become jaundiced. If in the absence of any other obvious cause for steadily increasing jaundice then other albeit uncommon possibilities should be sought. These include red cell enzyme studies (glucose 6 phosphate dehydrogenase and pyruvate kinase) where a deficiency of either of these enzymes will give a haemolytic picture, hypothyroidism, galactosaemia or Gilbert's syndrome. Gilbert's syndrome is seen in about 1 in 20 individuals. In this inherited condition there is diminished activity of the bilirubin converting enzyme (uridine diphosphate-glucuronyl transferase) in the liver. If there is an ABO incompatibility situation with accelerated destruction of red blood cells then bilirubin levels will be significantly raised in Gilbert's disease. Otherwise bilirubin levels are no different from other babies. Gilbert's disease is an autosomal recessive inherited condition meaning that each parent carries this recessive gene. Each baby born will therefore have a 1 in 4 chance of inheriting the condition by receiving the recessive gene from each parent.

Chapter 15 : The Breasts & breast feeding

The mammary glands (breasts) secrete milk and according to Oliver Wendell Holmes "a pair of substantial mammary glands has the advantage over the two hemispheres of the most learned professor's brains in the art of compounding a nutritious fluid for infants". Disappointingly and contrary to common expectations there is no correlation between the size of the breasts before pregnancy and the amount of milk that can be produced after the baby has been born.

Each breast is made up 15 – 20 lobes of glandular tissue which are connected by fibrous tissue and with fat occupying the intervals between the lobes. The lobes themselves are made up of lobules which when fully developed consist of clusters of secretory units called alveoli. These alveoli open into the smallest branches of the lactiferous ducts which eventually unite and end in excretory ducts which converge towards the alveolar of the breast where they form dilatations called lactiferous sinuses. The lactiferous sinuses act as a reservoir for the milk and they perforate the nipple with 15-20 orifices. Around the nipple is a coloured area of skin called the areolar which in the virgin state is a delicate rose colour. From the second month of pregnancy the areolar enlarges and takes on a darker colouring eventually becoming dark brown or black as the pregnancy progresses. The colour lightens once lactation is completed but does not disappear entirely. On the surface of the areolar are numerous fatty glands which become enlarged during lactation and take on the appearance of tubercles beneath the skin. They secrete a fatty substance which acts as a protection for the skin and the nipples. In a

Chapter 15 : The Breasts & breast feeding 119

lactating breast oil globules accumulate in the cells lining the secretory units (alveoli) and these oily droplets are discharged into the lactiferous ducts when the epithelial cells containing the oily droplets are ruptured. They rupture because each alveolus is surrounded by a layer of myoepithelial cells which can contract and when they do, milk is squeezed into the lactiferous ducts. When they are secreting milk the alveoli are under the influence of maternal prolactin but when suckling commences another hormone oxytocin is released and this causes the actual milk let down. Thus prolactin and oxytocin which are hormones from the pituitary gland are responsible for the synthesis of the milk and the let down respectively. From the sixth month of the pregnancy the nipples should be expressed regularly and daily by pressing on the areolar with the thumb and fingers trying to mimic the baby's sucking reflex. Lanoline can be massaged into the nipples at the same time. Even retracted nipples can be helped by this procedure. Colostrum can be expressed in the last month or two if this procedure is carried out daily and this will prevent painful distension of the breasts caused by inspissated colostrum which otherwise would have been retained in the lactiferous ducts and sinuses after the delivery.

Once baby is born physical contact and suckling should be encouraged as soon as possible – skin to skin – and frequent demand feeding should be allowed. The early ingestion of colostrum will serve to protect the baby against diarrhea and will also assist in the elimination of meconium by encouraging peristaltic movements of the bowel. The real milk comes in by the second to the fifth day and will come in sooner in the multiparous mother than in a first time mother.

When the newborn baby is first put to the breast sucking should only be allowed to continue for two to three minutes only on each side but as frequently as needed because prolonged feeding at this stage will cause excoriation and soreness of the nipples. After a few days the nipples will "harden"as a protective epithelial coating forms over them. The epithelium of the nipples is very delicate and can easily be damaged with prolonged suckling in the first few days. A good and sensible plan which I strongly suggest is initially to allow for 2-3 minutes of sucking on each breast as often as wished for the first 2-3 days and then increase by one minute for each breast each day. Breast feeding is usually well established by 7-10 days and should not last longer than 10-15 minutes on each side. Both breasts should be offered at each feed and each breast emptied.

No more than a minute or two for "winding" between each side is necessary. If too much time is spent on the first breast or there are any time wasting procedures between each side, baby may become disinterested or tired and not empty the second breast. This will eventually lead to breast engorgement and later mastitis or a breast abscess. When baby is put to the breast the nipples become erect and successful breast feeding will depend on the ability of the baby to stretch the nipple against the baby's hard palate (roof of the mouth). During sucking the baby's lips, gums and tongue create a vacuum in the oral cavity and with sucking a "clicking" sound may be heard as the vacuum is broken. Once baby has taken the nipple and started to suckle, after a minute or two the experienced breast feeding mother will feel a tingling sensation and retraction of the breasts which is due to the "let down" reflex and which signals the release of the "hind milk". Milk from the lactiferous sinuses passes from the nipple and is squirted into the baby's mouth being helped by peristaltic "chewing" movements by baby.

Between breast feeds milk secretion is occurring at a constant and uninterrupted rate under the influence of prolactin, a hormone secreted by the pituitary gland. This milk is low in fat content (2%) as well as protein and is emptied into the lactiferous sinuses where it awaits ingestion by the baby at the next feed. This passively and spontaneously secreted milk is the "fore milk" and makes up about one third of the milk volume available to baby. With the onset of sucking prolactin induces the formation of more "fore milk". Two or three minutes later the same sucking reflex causes the pituitary gland to release another hormone oxytocin which is responsible for the "let down" reflex resulting in the release of the "hind milk". This is a high calorie milk rich in fats (4%-7%) and which makes up the remaining two thirds of the milk available to the baby. The breast milk at the end of each feed contains four to five times as much fat and one and a half times as much protein as at the beginning of the feed. The carbohydrate (lactose) content is unaltered.

In summary an experienced breast feeding mother will find that when baby is put to the breast, the baby will suck forcefully for a few

Chapter 15 : The Breasts & breast feeding

minutes as the "fore milk" is taken. Baby will then move into a steady rhythmical sucking mode as the "hind milk" which has been stored in packets (vacuoles) is released and drunk. The feeding time for each breast should not need to be greater than 10-15 minutes and some babies are very efficient and may empty the breast in a shorter time. However complete emptying of both breasts at each feed is the key to successful breast feeding. The lactating mother may need to increase her caloric intake by 500 calories a day and her fluid intake by up to 1 litre per day. A well nourished mother should be able to meet her baby's nutritional requirements with breast milk alone for at least the first 4 months. There is plenty of water in breast milk and therefore there is no need for extra water to be given to the baby even in the summer and especially in a temperate climate. Breast milk is sufficient and also there are no calories in water. There is an average and normal weight loss of about 10% in purely breast fed babies in the first few days of life and this has no detrimental effect on the baby's health. Over the first 10-15 days milk requirements gradually increase to about 150mls/kg/day (2.5ozs/lb/day) which will meet all the nutritional needs of the baby.

The World Health Organization defines exclusive breast feeding to mean breast feeding only and not including water or juices. In the Western world women who breast feed baby beyond 6 months are considered to be prolonged or extended breast feeders and breast feeding after 12 months reflects a maternal philosophy rather than a form of nourishment. In some countries however the average duration of breast feeding is 2-3 years. Extended breast feeding is associated with a 30% reduction in the risk of pre-menopausal breast cancer. Up until the last century or two breast feeding as well as baby and mother sleeping in the same bed were usual but present Western values have tended to favour early weaning as well as solitary infant sleeping habits. The number and duration of breast feeds will be increased in the bed sharing situation. Remember that bed sharing can result in overlaying or suffocation. Babies who routinely sleep in mother's bed will feed three times more often during the night than babies who sleep separately. By sleeping in close proximity to mother the infant receives olfactory cues from mother and mother is also more likely to have a better sense of the baby's subtle sounds and movements during bed sharing. Newborn babies respond to olfactory differences between washed and unwashed breasts because

Chapter 15 : The Breasts & breast feeding

as well as secreting colostrum and milk the nipples and areolar are rich in glands which exude attractive odours. These naturally occurring odours have a role in guiding baby to the nipple and unnecessary routine cleaning of the nipple will reduce or eliminate these cues. Apparently baby girls but not baby boys prefer the lactating mother's breasts over clean breast pads. Baby girls recognize and prefer the breast odours of their own mother suggesting that lactating women produce chemical substances that newborn girls find attractive. These findings do not apply to the newborn male. The organs of sucking and articulation are the same and in males it has been found that there is relationship between breast feeding and clarity of speech while there is improved tonal quality in both sexes.

To enhance breast feeding mothers may eat garlic since babies will feed for longer and suck more if human milk is flavoured with garlic. A glass of beer (or low alcohol beer) once or twice daily will increase milk production by increasing prolactin levels. Drugs such as Chlorpromazine or Metaclopramide given to mother will stimulate the milk supply in women with low prolactin levels especially after a premature delivery. However the optimum stimulus for prolactin release is a sucking stimulus every 2-3 hours. Inhibitory factors which will cause milk insufficiency include pain, anxiety and fatigue. If there is an insufficient supply of milk there will be an absence of breast distension and poor weight gains by baby. If breast distension is absent, milk leakage or over abundant milk leakage at night as well as milk ejection at the end of a feed will usually mean inappropriate and excessive oxytocin release. Maximal physical exercise by the lactating mother will cause a significant increase in lactic acid concentration in breast milk which will affect the taste of the milk.

Lactic acid will remain in the milk for about 90 minutes and will give the milk a sour taste which can be detected by baby and may cause some resistance to feeding .

Breast milk contains less casein and more whey protein than cow's milk. Whey protein is the protein that remains in solution after the precipitation of casein from milk by the addition of rennin or acid and it is more easily digested by the baby than casein. Whey protein makes up about 70% of the total protein in breast milk, but only about 20% in cow's milk. More gastric acid (stomach juice) is

Chapter 15 : The Breasts & breast feeding

needed to curdle cow's milk but even so the stomach contents may be less acidic than in those babies who are breast fed. Decreased stomach acidity will reduce the bactericidal activity of gastric juice and this is one of the reasons why breast fed babies have fewer gastro-intestinal infections than cow's milk fed infants. Fat provides about 50% of the calories in breast milk and constitutes the major source of energy for the infant. The fat content of breast milk shows diurnal variations being highest in the early morning. The fat content of "hind milk" is 2-3 times greater than in "fore milk" and is the milk collected by complete breast emptying. Fat is an essential nutrient for myelination and development of the nervous system and the polyunsaturated fats (arachidonic and docosohexanoic acids) present in breast milk are essential components of the structural fats in brain and nervous tissue. The frequency of neurological abnormalities is said to be half as high in babies who have been breast fed exclusively for at least 3 weeks. Claims are also made for breast fed babies to show a few points increase in IQ later in life.

Breast milk is unique in its ability to enhance iron absorption and is sufficient to ensure an adequate iron balance in a full term baby for at least the first 6 months if solely breast fed. However by 9 months of age 50% of breast fed babies will show biochemical evidence of iron deficiency unless exogenous sources of iron are made available. Giving iron to the breast feeding mother will not change the iron content of breast milk.

Breast feeding and drugs

Almost all drugs ingested by mother will appear in breast milk although as a rule the amounts are small – in the order of 1%-2% of the maternal dose and do not affect the baby. However drugs that enter the infant's circulation in amounts greater than 10% of maternal levels should be avoided.

Points to consider

- does mother need the drug? - is there an alternative?
- take the drug immediately after breast feed has been completed
- drug levels can be measured in baby's blood if necessary.

Safe drugs include the following

- **Anticoagulants** - heparin, warfarin
- **Anti inflammatories** – diclofenac, indomethacin, ibuprofen, ketoprofen, naproxen and sulindac.
- **Steroids**.
- **Diuretics**
- **Anticonvulsants**.
- **Antibiotics** (except chloramphenicol, tetracyclines, ciprofloxacin, norfloxacin)
- **Antifungals** (except tinidazole)
- **Antiviral agents**.
- **Anti anxiety & psychotropics** (except lithium)
- **Thyroid agents** - propyl/methylthiouracil, thyroxine.
- **Cough mixtures**.
- **Caffeine** - coffee, tea, soft drinks. One cup of coffee (100mgm caffeine) provides 1mgm of caffeine to baby. One cup of tea (contains 50mgm of caffeine), one 12oz glass of coke (contains 50mgm of caffeine)
- **Alcohol** - this is transported rapidly to breast milk and the ratio of alcohol in maternal blood to milk is 1:1. Also the rate of disappearance of alcohol from maternal blood is the same as from breast milk.

Drugs that should not be taken if breast feeding

- **Anti neoplastic** (chemotherapeutic) agents and immunosuppressives.
- **Gold**.
- **Amphetamines**
- **Cocaine, heroin, marijuana**.
- **Radio-isotopes** - radioactive iodine stays in breast milk for 7-10 days. Gallium remains in milk for 2 weeks. Technetium remains in milk for 24 hours. Therefore interrupt breast feeding in each case until milk is clear.
- **Flagyl** (metronidazole) – stop breast feeding for 24 hours after a single dose.
- **Aspirin** – only use if there is no alternative.

Chapter 16 : Vomiting, posseting and spilling

Gastro-oesophageal reflux

Gastro-oesophageal reflux (GOR) is the involuntary passage of gastric contents from the stomach into the oesophagus. It is so common in babies that it is seen as a normal event and is then called spilling or posseting. GOR becomes pathological when complications such as oesophagitis (heartburn) causing apparent chest pain, irritability and writhing movements of the neck or trunk occur. Writhing movements (dystonic posturing) of the neck were known in the past as Sandifer's syndrome. Reflux of acidic stomach contents may cause inflammation of the oesophagus or the mucous membranes of the upper airways. Inflammation of the upper airways can cause laryngospasm, hoarseness due to acid erosion of the vocal cords, bronchospasm or apnoea. Recurrent aspiration pneumonia is possible and gastro-oesophageal reflux should always be considered in any baby who has had two or more bouts of pneumonia. Failure to thrive is occasionally a symptom.

An episode of GOR is defined as a fall in the lower oesophagus pH values to 4 or less for at least 8 seconds. The gold standard for the diagnosis of GOR is prolonged lower oesophageal pH values over a period of twenty four hours monitoring. However as a rule a Barium swallow is more commonly used to confirm the diagnosis since it is more practical. The reflux index is the percentage time during the study when the pH is less than 4. Percentage GOR times of up to

Chapter 16 : Vomiting, posseting and spilling

20% can be controlled by medical treatment but if these times exceed 30% then surgery may eventually be necessary.

Stomach contents are acidic and contain pepsin both of which irritate tissues which do not have defensive mechanisms against their corrosive effects. When milk is drunk it is normally retained in the stomach because the lower oesophageal sphincter becomes closed. The upper oesophageal sphincter when closed prevents air from entering the oesophagus. Swallowing causes both sphincters to open then close in sequence but complete closure may not be effected. The regurgitation of small amounts of milk without any other symptoms is so common that it is normal and the average newborn infant may reflux up to twenty times a day. Usually there are an increased number of reflux episodes in the awake state but during sleep the reflux episodes last longer. GOR is more frequent when the stomach is full but oesophageal clearing is faster in the awake and alert state. In the prone, right lateral or upright position the rate of stomach emptying is accelerated by gravity infusion into the duodenum and therefore air ("wind") rather than stomach juices will tend to reflux. Infants who sleep in the prone or right lateral position because of accelerated stomach emptying tend to sleep longer and cry less often than if in the currently recommended supine position. They have a slower heart rate and a more regular respiratory pattern. Most GOR will have remitted by ten months of age as oesophogeal sphincter maturation occurs.

Management of gastro-oesophageal reflux

- **propping** (positioning) to enlist the help of gravity where the baby is propped at about 30 degrees in a straight line from the horizontal by placing books or bricks under the legs at the head of the cot. To lessen the refluxing tendency baby is placed in the right lateral position with the lower arm well in front of the body to prevent rolling to the prone position. Otherwise the supine position (on the back) is used because the prone position is considered to be an independent risk factor for sudden infant death syndrome. Rolled up blankets or wedges should not be used to support the right lateral position because baby may roll over. An infant seat with a

Chapter 16 : Vomiting, posseting and spilling   127

sixty degree angle may not help since this upright position places the gastro-oesophageal junction under a pool of gastric contents. The oesophagus enters the stomach posteriorly and when the lower oesophageal sphincter undergoes intermittent normal and spontaneous relaxation, stomach contents covering the gastro-oesophageal junction will be regurgitated into the oesophagus.

- **Milk thickeners** such as Gaviscon (sodium alginate and magnesium alginate) increase the viscosity of the milk and form a thick foamy raft on the surface of the gastric contents. The thickened stomach contents will reflux less readily. Gaviscon will also coat the lining of the oesophagus giving it a protective coating thus lessening the inflammatory effect of regurgitated acidic stomach contents. Gaviscon lessens but does not prevent gastro-oesophageal reflux. If baby is breast fed Gaviscon must be given before the feed and mixed into a creamy paste with expressed breast milk. The paste is spooned in and then followed by the breast feed. If baby is bottle fed Gaviscon is mixed into a creamy paste again and then put into the bottle and the bottle well shaken. Gaviscon will tend to thicken or firm the bowel motions. Other food thickeners are Karicare food thickener (pregelatinized maize starch), Carobel a thickening agent made from carob seed flower, Nestargel and already made up thickened formulae. In these formulae 30% of the lactose in the formula is substituted with pre-gelatinized amylopectin rice starch (2.3gms starch/100mls). This maintains the appearance, osmolality, caloric value and nutrient profile of the routine formula. It provides greater viscosity which increases further with exposure to stomach acid.

- **Ranitidine** (Zantac), a histamine H2 receptor antagonist in a dose of 2mgm/kg/day or Omeprazole (Losec)) a proton pump inhibitor in a dose of 2mgm/kg 12 hourly. Both these agents diminish the volume of acid secreted by the stomach. The decreased acidity of stomach secretions means a less corrosive and irritating effect on an inflamed oesophagus.

- **Cisapride** (Prepulsid) and Metoclopramide (Maxolon) are pro-kinetic agents which increase gastro-intestinal motility and stomach emptying. Cisapride is now contra- indicated because it may be responsible for long QT syndrome which has been implicated in some cases of sudden cardiac arrest. Metoclopramide can be associated with dystonic side effects if used for prolonged periods of time. These agents are best avoided if possible.

If full medical treatment has failed or an oesophageal stricture is forming then surgery (Nissen fundoplication) may become necessary.

In silent GOR there may be little or no vomiting but symptoms in the way of coughing, choking, gagging, swallowing movements (ruminating) or breath holding may be a feature. Any or all of these symptoms represent an attempt by the baby to avoid aspirating regurgitated stomach contents into the lungs. If oesophagitis (heartburn) due to regurgitated acid stomach contents is present then writhing movements of the neck or trunk together with a distressed or unsettled baby will become obvious. GOR should be considered in any baby who presents in any of the following ways – apnoeic (breath holding) spells, recurrent pneumonia, bronchitis, recurrent croup, iron deficiency anaemia, failure to thrive, abnormal posturing of the neck or trunk, rumination with restlessness and apparent discomfort. In older infants dental erosion due to loss of dental enamel by regurgitated acid stomach contents can occur. For this to happen there must be a pH value of less than four for at least 25% of each 24 hour period. Because the pH of gastric contents is less than one, the local buffering agents in the mouth will be overwhelmed leading to surface enamel dissolution. The severity of this damage will range from white spots on the teeth to a flattening of the cusps, eventual dentine exposure and infection.

Pyloric stenosis

The pyloric orifice is the opening through which the stomach communicates with the duodenum. Hypertrophy (thickening) of the

Chapter 16 : Vomiting, posseting and spilling 129

pyloric musculature (sphincter) leads to the obstruction of stomach contents leaving the stomach. The presenting symptom is vomiting which is not bile stained and which gradually becomes more marked and frequent and eventually projectile in nature. Typically vomiting occurs during a feed or shortly after and is accompanied by writhing squirming movements of the baby before the projectile vomit. Vomiting due to pyloric stenosis can occur in the first week but the usual time of presentation is between the 4th and 6th week. Poor or no weight gains become apparent and baby is hungry and dehydrated. If baby is examined during a feed when the abdomen is soft and relaxed and with appropriate oblique lighting visible peristaltic movements will be seen passing across the abdomen from the upper left quadrant to the lower right as the stomach contracts against the obstruction. Usually with much practice an olive shaped "tumour" at that time will be palpable. Baby will then have a projectile vomit with no bile present. If the pyloric "tumour" is ripe an ultrasound study of the pyloric region will confirm the diagnosis by demonstrating a thickened pylorus.

Pyloric stenosis is inherited as a polygenic dominant trait and affects about 2-3/1000 live births with males being affected 4-6 times as often as females and is more likely in the first born male. Pyloric stenosis develops in about 20% of the sons and 7% of the daughters of mothers who have had pyloric stenosis. If the father had pyloric stenosis then 5% of his sons and 2.5% of daughters will be affected.

These figures suggest that in the female patient there is a stronger genetic influence over the condition. Treatment is surgical and the surgical approach was described over 80 years ago by Conrad Ramstedt a Prussian military surgeon and it has been preserved more or less intact over the years. Prior to surgical division of the pyloric "tumour" slow starvation and death was the inevitable outcome for the baby.

Chapter 17 : Milk protein sensitivities

Colic (infantile colic, evening colic, three months colic)

This is an ill defined condition characterized by the sudden onset of severe abdominal pain accompanied by paroxysmal screaming, stiffening of the body, drawing up of the legs or arching of the back. The abdomen feels hard and may be distended and there are usually frequent loud bowel sounds which are accompanied by the passage of considerable flatus. During these attacks the infant is inconsolable. The attacks come in waves and fluctuate in severity and can last several hours. They occur typically in the late afternoon or early evening and often show a rhythmical and almost predictable pattern during the twenty four hour period. Colic has usually disappeared by three to four months but can last beyond that age. With milder attacks the baby may merely be uncomfortable or unsettled especially in the evenings.

Colic occurring after each feed but not necessarily in the evening is related to parental smoking and the chance of colic occurring is doubled if mother smokes fifteen or more cigarettes daily. Maternal smoking during the pregnancy also increases the risk of colic. Colicky babies are in all other respects perfectly normal. The incidence of colic is about 13% in all babies and invariably there is a family history of allergies in terms of asthma, hay fever or eczema. Subsequent exposure directly (cow's milk formula) or indirectly (mother's diet and breast milk) of these infants to cow's milk products, nut or egg protein triggers colic. The symptoms of colic are consistent with paroxysmal painful spasm of smooth muscle in the

Chapter 17 : Milk protein sensitivities

large bowel which will come in waves. The definition of colic may vary widely. For instance if baby has screaming paroxysmal attacks of pain for more than three hours each day, occurring on at least three days a week and over a period of three weeks or more then by these strict criteria the incidence of colic is about 3%-4%. If more liberal criteria such as sudden onset of screaming, stiffening or arching of the back and where baby is inconsolable for minutes to hours, then the incidence is about 13%.

There are no differences otherwise in temperament or sleeping patterns between babies with or without colic. There may however be an association between infant colic and mother's perception of baby's temperament and sleeping pattern and this perception can cause a functional interaction between mother and baby caused by a combination of maternal fatigue, guilt, depression or resentment. Even after colic has disappeared the relationships between parents and baby may reflect early unhappy parenting experiences. Parents need to know that colic is temporary and is not the fault of anybody and hopefully not be an event that will lead to a sustained negative view of the infant. Nearly sixty years ago it was suggested that food antigens transmitted in mother's breast milk could cause colic and later cow's milk protein in breast milk was found to be the chief offender

If a breast feeding mother completely eliminates all cow's milk protein products (milk, butter, cheese, yogurt, ice cream and bread with a milk powder base) together with nut and egg protein from her diet for at least twelve days then colic is usually relieved.

If not then consider other less likely possibilities such as fish, chicken, honey or citrus fruits. The diagnosis is confirmed by the return of colic if these proteins are abruptly re-introduced after this time. Gradually the infant will develop a tolerance to these antigens. A family history (siblings or parents) of allergies such as asthma, eczema or hay fever is usual.

As a rule colic will begin at about two to four weeks and last three or four months.There have been many treatments offered for colic and some of these have seemed to be of help for desperate parents. Dicyclomine syrup an antispasmodic agent if given half an hour before the expected attack of colic could limit the severity or prevent the attack. However it has been withdrawn from the treatment of colic because of the suggestion that it caused apnoeic (cessation of breathing) attacks in the baby. Simethicone is a defoaming agent which reduces the surface tension of gas bubbles in the bowel causing them to coalesce and thus accelerate the passage of gas through the bowel but it has no effect on painful muscle spasm which is the problem. Sleep tight is an electro-mechanical device which if attached to baby's cot will produce continuous vibrations. It is meant to mimic the auditory and movement effect of riding in a car traveling at a reasonable speed. Certain other sounds associated with vibration such as motors, vacuum cleaners, clothes driers or washing machines will sometimes give relief. Gripe water (dill water) was an old fashioned remedy and was a dill flavoured sugar solution containing alcohol, but the modern preparation no longer contains alcohol. However a 12% sucrose (sugar) solution given in a dose of 2mls immediately the attack starts can relieve colic in a few minutes in about 50% of babies. It is safe and harmless and the effect is due to the release of endorphins (which have an opiate like effect) from the infant's brain. In the past chewed dates (70% sugar as glucose and fructose) placed in baby's mouth would give relief from the pain of circumcision presumably for the same reason. Cromoglycate (Nalcrom) in a dose of 50mgm - 100mgm in a 2.5mls solution given prior to each feed can be very effective in reducing or abolishing intractable colic and an additional virtue is that it has no side effects. Hydrolysed formulae in which the milk protein is partially fragmented, extensively fragmented or completely broken down into elemental amino acid components of the protein can cure colic. A whey hydrolysate formula is more effective in relieving colic than a casein hydrolysate formula or soy or goat's milk formulae.

Cow's milk protein allergy or intolerance

There is a distinction to be drawn between an allergic reaction to a food product and an intolerance (sensitivity) to that food. A food allergy is an immunological (IgE mediated) reaction to one or more food proteins which are usually cow's milk and its products, peanuts and egg white.

A food intolerance (sensitivity) is the term used for non-immune reactions to foods and can be responsible for colic, a generally

Chapter 17 : Milk protein sensitivities

unsettled baby or for an exacerbation of vomiting in gastro-oesophageal reflux. There is no definitive simple test to establish conclusively a diagnosis of food allergy or intolerance but in both these conditions a family history of similar problems in the past may be elicited.

A broader and more acceptable term for food allergy or intolerance would be food sensitivity and this would encompass the various reactions to different food allergens.

The prevalence of cow's milk protein sensitivity is up to 7% of infants and depends on the criteria used for the diagnosis. Allergic symptoms to cow's milk protein products have usually disappeared in about 85% of infants by two years of age and in 95% of children by six years. Cow's milk sensitivity is more common in atopic families where there is a history of urticaria, eczema, asthma or hay fever. Food sensitivity reactions may be immediate or delayed. In the immediate reaction an urticarial rash will usually appear within 45 minutes or more dramatically and rapidly an anaphylactic reaction with shock although this is rare. With the delayed response there can be many and often apparently unconnected symptoms such as an exacerbation of eczema, colic, vomiting, diarrhea which may be blood stained, constipation, coughing, failure to thrive with listlessness or a generally unsettled state. Most infants with the immediate reaction response will show a positive skin prick result with the offending agent, while 75% of infants with a delayed response will be negative for the skin prick and intra dermal tests. Blood tests may show an elevated eosinophil count or IgE levels in those with symptoms who are ingesting cow's milk protein either directly from formula or indirectly through breast milk.

Cow's milk protein can be broken down into four components, namely casein (28mgm/ml); beta-lactoglobulin (4mgm/ml); alpha-lactalbumin (0.7mgm/ml); and bovine serum albumin (0.2mgm/ml). While casein has the highest concentration it is not considered to be as allergenic as the other cow's milk proteins. The total protein concentration in cow's milk is 3.3gms/litre and the two main components are casein (insoluble curd) making up 80% and whey protein which is soluble the remaining 20%.

Chapter 17 : Milk protein sensitivities

Whey protein consists of three proteins, beta-lactoglobulin, alpha-lactalbumin and bovine serum albumin. The four major bovine casein groups are divided into alpha, beta, gamma and kappa. The major whey protein is beta-lactoglobulin and there are three genetic variants (A,B and C) and a single cow may produce any of these three variants either singly or in combination. Whey proteins are hydrolysed at a considerably slower rate taking longer to digest than casein. In addition cow's milk whey protein hydrolysates have a more pronounced allergenic property than casein hydrolysates since beta-lactoglobulin in whey protein is considered to be the most allergenic of all the cow's milk proteins. While these protein hydrolysate formulae are less likely to cause allergic symptoms than conventional cow's milk formulae, they still contain partially fractionated protein, the products of beta-lactoglobulin digestion (peptides) and can retain allergenic activity. In an elemental formula (Neocate) any animal or vegetable protein is completely replaced by synthetic individual amino acids (the building blocks of proteins) and as a result is more hypoallergenic than any other formula. The antigenic effect of cow's milk is not completely destroyed by boiling. Casein is heat stable but bovine serum albumin is heat labile, and beta-lactoglobulin and alpha-lactalbumin are of intermediate lability and thus are not completely denatured by boiling. High levels of beta-lactoglobulin the major milk protein allergen may be found in maternal breast milk and be responsible for the various symptoms of milk sensitivity. Similarly ovalbumin from egg white may also be found in breast milk and this protein can cause similar symptoms.

It is suggested that the hereditary risk of allergies can be nullified by six months of exclusive breast feeding but I have yet to be convinced that this practice while commendable is effective. The most common allergenic foods in the maternal diet are fish, citrus, fruits, tomatoes, honey, nuts, eggs and cow's milk protein products. Drinking cow's milk can cause normal infants to lose occult blood in the bowel motions (cow's milk protein colitis) and eventually develop an iron deficiency anaemia which can be associated with cognitive handicaps. Therefore full strength cow's milk should be avoided for at least the first year of life and for this reason modified conventional baby milk formulae are fortified with iron. This leakage of occult blood in the bowel motions can be prevented by the substitution of soya bean formula, goat's milk formula or a protein hydrolysate formula. The treatment of iron deficiency anaemia with iron in these situations without modifying the formula will not stop the loss of occult blood.

Carbohydrates

The most common carbohydrates in the infant's diet are lactose and sucrose both of which are double sugars (disaccharides) as well as amylose and amylopectin which are polysaccharides. Lactose is the main carbohydrate in milk and when ingested is broken down into two monosaccharides glucose and galactose. Sucrose is broken down into the monosaccharides glucose and fructose, and amylose and amylopectin are large glucose polymers that are found in corn syrup. The nature of the sugars is important because specific enzymes which reside in the brush border of the mucous membrane lining the villi (finger-like projections) of the small bowel are responsible for breaking down the various carbohydrates so that they can be absorbed. There are three enzymes (called disaccharidases) which hydrolyse the various carbohydrates. These enzymes are lactase, sucrase and glucoamylase. Sucrase and glucoamylase hydrolyse maltose into glucose and can be called maltases and their maximal activity is near the mid point of the small bowel villi. On the other hand lactase converts lactose (milk sugar) into glucose and galactose and its maximal activity is found at the tip of the villi. Lactase levels are also lower than maltase levels and this fact coupled with the site of activity at the tip if the small bowel villi makes lactase more susceptible to mucosal damage resulting in diminished lactase activity. Thus diarrhea will be exacerbated by lactose ingestion if there has been mucosal damage such as in gastro-enteritis. Because there are lower levels of lactase than maltase the hydrolysis of lactose is about half as fast when compared with other carbohydrates and it is this fact rather than the transport of the hydrolytic products glucose and galactose across the small bowel that is the rate limiting factor for the assimilation of lactose. With a secondary disaccharidase deficiency as a result of gastro-enteritis unabsorbed carbohydrates remain in the bowel and increase the osmotic load of the contents of the bowel causing water to move into the bowel lumen. Intestinal motility increases and the partly digested carbohydrates are fermented by intestinal flora to form organic acids, carbon dioxide and hydrogen. This then causes abdominal bloating and cramping and frothy diarrhea. Infants recovering from acute gastro-enteritis tolerate a soy formula containing sucrose and corn syrup better than a lactose containing formula. The soy protein formula contains two carbohydrates that are digested and absorbed by separate pathways and have an advantage over formulae containing only one carbohydrate.

Chapter 18 : Some symptoms and signs

Bowel motions

Breast fed babies if they are well fed will have motions ranging from frequent frothy explosive yellow motions which are due to relative lactose intolerance and may contain white pellets of calcium soaps, or a soft brown motion up to every ten to twelve days with all other variations in between. The motions can vary in consistency from soup like, unformed cow pats, soft toothpaste or sausage like, to putty and finally hard pellets or pebbles. Pellet or pebble like motions mean constipation no matter how frequently or infrequently they are passed. Constipation in a baby usually reflects insufficient fluid. On the other hand a soft motion every few days is normal and is not constipation.

The bowel motions are green in babies when an iron fortified whey dominant formula is used. Those babies receiving a whey dominant low iron formula will have yellow motions. Babies receiving a casein dominant iron fortified formula will have yellow or brown motions although they can occasionally be green. Soft bowel motions are seen in babies taking either a whey or casein dominant formula but watery or slimy motions are more common with a whey dominant formula. The changes in the colour of the motions are probably due to the changes in the bowel micro-flora associated with the various formulae.

Hard motions will stretch the anus and cause an anal tear. This then makes the passage of further motions painful leading to

Chapter 18 : Some symptoms and signs   138

retention and an exacerbation of constipation. If there is an anal fissure or tear, the outside of the hard motion may be streaked with blood. But remember hard infrequent motions in the presence of abdominal distension should suggest a bowel obstruction such as Hirschsprung's disease and not be ascribed simply to constipation.

White motions in a new jaundiced baby mean neonatal hepatitis or biliary atresia. In these two conditions the bile ducts are blocked preventing bile from reaching the bowel to colour the motions.

Urgent medical assessment is needed for white motions or for constipation with obvious abdominal distension.

Breast tissue

Breast enlargement may persist normally in baby for several months. "Witch's milk" can be expressed from the nipple if the breasts are squeezed. Squeezing is usually accidental when baby is clasped around the chest when lifted and held. Excessive manipulation of baby's breast tissue can lead to inflammation or an abscess.

Occasionally in a female toddler there will be early unilateral growth of a breast (premature thelarche) and this can be quite normal provided certain endocrinological screening tests are normal also.

Breathlessness

If accompanied by coughing and wheezing in a baby then it will be usually be due to bronchiolitis. Similar symptoms occur with asthma but wheezing in asthma is unlikely before nine months of age.

A poor or dusky colour and poor weight gains may mean heart failure in which case breathlessness will be worsened during feeding because of the added exercise brought about by feeding.

Fever with very rapid shallow breathing and with or without a cough but not necessarily any wheezing will suggest pneumonia which may be bacterial or viral in origin.

Chapter 18 : Some symptoms and signs 139

Cheeks

Unilateral flushing (hectic spot) of the cheek probably has teething as a cause. Bilateral flushing of the cheeks usually means an elevated temperature regardless of the underlying cause.

Bright red cheeks (slapped cheek) is usually the result of fifth disease (slapped cheek syndrome), a parvo virus infection.

Coughing

A series of paroxysmal coughs followed by an inspiratory "whoop" is due to whooping cough and in a small baby usually means a hospital admission for observation and if necessary treatment. Paroxysmal coughing may also be a feature of bronchiolitis when there is a frequent harsh repetitive cough with rapid breathing and wheezing. If baby can still feed comfortably then baby can probably managed at home. If there is difficulty with feeding then a hospital admission is warranted.

Single brief coughing episodes as though trying to clear the throat in association with spilling or gagging are characteristic of gastro-oesophageal reflux.

A cough accompanied by a "runny" nose is usually due to an upper respiratory tract infection and is usually viral in origin.

Dehydration

The skin will be tented and dry when pinched or lifted because of loss of elasticity and the anterior fontanelle will be sunken. If there is associated diarrhoea, gastro-enteritis is the cause. If on the other hand in an older infant there is the passage of excessive and copious urine (polyuria) and there is excessive thirst round the clock diabetes mellitus is likely and should mean checking the urine for glucose and an early blood glucose level.

Dehydration whatever the cause requires prompt medical assessment.

Chapter 18 : Some symptoms and signs

Diarrhoea

A normal healthy breast fed baby will have frothy yellow motions which may contain small white pellets of calcium soaps.

With gastro-enteritis the motions will be frequent and mucousy – flecks of blood will be mixed in the motions if it is bacterial gastro-enteritis but no blood will be seen in the more common viral gastro-enteritis. In the absence of gastro-enteritis and with the presence of blood in loose yellow motions consider cow's milk protein colitis.

Dummies (pacifiers, soothers, binkies)

Every baby has a strong need for sucking especially in the first six months of life. Beyond this time the use of a dummy becomes more of a habit perhaps producing a feeling of security. After six months while it may seem to be a harmless enough habit it can lead to minor dental malocclusions which will however regress spontaneously and is not as harmful to teeth alignment as is thumb sucking. Prolonged use of the dummy though will lead will lead to malocclusions particularly anterior overbite. Dummies have also been implicated in an increased incidence of middle ear infections. The dummy should be discarded completely by at least ten months of age and preferably earlier if possible.

The use of the dummy may shorten the period of breast feeding and have some bearing on weaning. There may be reduced motivation to breast feed or it may be a marker for an infant who is having difficulty sucking at the breast. It may be perceived as a reason for baby getting insufficient breast milk. The use of a dummy for more than two hours a day is more common in babies of mothers who do not know how to correctly breast feed.

However dummies can aid in sleeping, can be comforting and if coated in sugar may relieve pain. Specially designed dummies have been made to imitate the shape of mother's breast as it is when inside baby's mouth during feeding. These dummies are said to allow for a more natural arch development because they encourage muscular movements that closely resemble those seen with breast feeding. Nevertheless no clinically significant differences in the development of dental problems have been seen in children who have used these specially designed dummies compared with ordinary dummies. In brief dummies are acceptable for the first six months of life and should then be discarded before they become a habit with undesirable consequences.

Ears

Tugging or pulling the ear in association with an elevated temperature and a flushed cheek may mean an ear infection (otitis media).

Tugging the ear in a well non-febrile but unhappy infant with a red cheek (hectic spot) is probably due to teething.

A discharge from the ear may be waxy, clear, bloody or purulent and if the temperature is raised consider an ear infection.

Eyes

Persistent or intermittent deviation of one or both eyes after 4 months needs attention to exclude a squint (strabismus). Babies have prominent epicanthic folds at the inner corner of the eyes and this inner fold of the upper eyelid will cause a spurious squint to become apparent when baby looks to one side or the other. The white part (sclera) of the eye will disappear behind the epicanthic fold and it will seem that the eye has turned inwards. A cover test when a brightly coloured object is held in front of the baby and one of the eyes covered will determine if the squint is spurious or real. Baby is held on a parent's knee and facing outwards. The other parent holds the coloured object in front of baby thus engaging baby's attention and with the other hand covers one of the baby's eyes. Once the object is fixed the occluding hand is quickly removed. If there is a true squint the eye which has been covered will flick back to the midline from a deviated position while if the squint was spurious due to a prominent epicanthic fold there will be no deviation of the uncovered eye which will remain straight.

Fontanelle

The anterior fontanelle is normally soft and slightly depressed. The fontanelle can only be properly assessed when baby is awake, settled and in a sitting or vertical position. If under these conditions the fontanelle is full, tense or bulging then meningitis, a brain tumour or heart failure must be considered.

A dehydrated baby will have a sunken fontanelle and the skin will be dry or wrinkled with loss of normal tissue elasticity.

Jaundice

Jaundice can persist in a healthy breast fed baby for several weeks and provided the motions remain yellow and are not white and the urine is clear and not dark brown, then there is no cause for alarm and breast milk jaundice is the usual reason.

White motions with dark urine mean an obstructive form of jaundice (biliary atresia or neonatal hepatitis) and needs urgent investigations.

With persisting jaundice and normal coloured but very hard motions (constipation) along with cold extremities consider low thyroid function (congenital hypothyroidism).

Length

There is a 50% increase over the birth length by 15 months. As a rule the height of an adult will be twice that of the child who is just over 2 years of age.

Lethargy

If accompanied by a full fontanelle (with baby in a vertical position) and with or without a petechial rash then meningitis must be excluded urgently.

If febrile and with smelly (fishy or stale) urine then a urinary tract infection is probable.

Chapter 18 : Some symptoms and signs

Rash

A faint speckled rash on the trunk, face or limbs with or without a fever is usually indicative of a viral illness. A high fever for several days in an apparently well baby followed by the appearance of a fine rash when the fever subsides is likely to be "baby measles" (roseola infantum: exanthem subitum). In this case the occipital nodes will be enlarged and easily felt. The same rash but with fever and large occipital nodes may be german measles (rubella).

A fine rash over the chest or abdomen in a well baby could be due to detergents or soaps used in washing the baby's clothing. This sort of rash is called contact dermatitis.

A widespread petechial or purpuric rash could be due to sepsis and an urgent medical assessment would be needed.

Sleeping

Just over 50% of babies have their longest sleep time between 11pm and 7am. This may be because mother is likely to be asleep also at this time and would not be woken if baby was awake but quiet. New babies sleep between 1-5 hours at a time but by 3 months the sleep pattern is more predictable and they may begin to miss out one evening feed. Babies will usually sleep for 15-18 hours out of 24 each day. By one year the sleep time has reduced to about 10-12 hours. The explanation for the periodicity of sleeping and waking times is not known although most people feel that hunger is the main driving reason. Spontaneous jerking or startle movements (sleep jactitations) occur more often during deep sleep and are not necessarily in response to any stimulus. Babies have rhythmical sleeping patterns alternating between profound (quiet) sleep and light (active) sleep. Babies should sleep on their backs (supine) because of the possibility of cot death and not on their side or front (prone). If babies are put to sleep on their side only about 30% will be in that position in the morning, most will be supine and a few on their front. The head should be uncovered when asleep and loose clothing or bedding should be avoided. On the other hand baby should not be too tightly swaddled. If baby is sleeping in parents' bed be aware of the possibility of smothering due to overlaying and also the dangers of long hair strangulation. Babies should of course be living in a smoke free environment. The room temperature should be comfortably neutral and not too hot. About 17-20 degrees C is sufficient.

Smoking

The pregnant mother who smokes, smokes for two. Maternal smoking is associated with an increased risk of prematurity, growth retardation, spontaneous abortion and sudden infant death (cot death) syndrome and later respiratory disease for the infant. Pregnant women who smoke passively (absorbing smoke from the environment), let alone actively, accumulate measurable levels of nicotine and its metabolite cotinine in the blood as does the baby. Cotinine levels in blood, urine or saliva give an accurate record of exposure to cigarette smoke. Cotinine is incorporated into the growing hair shaft of the neonate and thus provides an historical record of exposure to cigarette smoke during foetal development.

Speech

This is known as expressive development as opposed to comprehension (receptive development) which manifests itself earlier. Babies can therefore understand more than they can express.

Expect throaty cooing sounds at about 6-8 weeks and guttural vocalising by 3-5 months when at this age baby will begin to have prolonged "conversations" with mother. Single syllables are heard from 5-7 months and double syllables by 7-9 months. The sequence of syllable development is guttural sounds (ga ga) first, followed by lingual (tongue) sounds (da da), then labial (lips) syllables (mum mum). A few words with meaning will be heard by 12 months, 2-3 word phrases by 18 months and short sentences by 2 years.

Teething

Teeth usually begin to appear from about 6 months and primary teeth continue to erupt until 6 years. The lower central incisors appear first. Teething does not cause any illness but may be associated with a red cheek (hectic spot), pulling of the ear and excessive dribbling or drooling. Rubbing the gums with a mild antihistamine syrup which baby can swallow safely can be helpful in alleviating any distress.

Temperature

The usual normal temperature range lies between 36.5-37.5 degrees centigrade. Persistent fever (38.5- 40 degrees centigrade) in an unwell baby should mean a search for a focus of infection. Conditions such as meningitis, ear infection, pharyngitis, upper or lower respiratory tract infection, gastro-enteritis, urinary tract infection, osteomyelitis or septic arthritis should be specifically looked for and excluded

If febrile consider any form of sepsis and remember lethargy with fever requires an urgent medical assessment.

Urine

The urine is a clear or light straw colour in the first two to three days of life and may leave a pink rose colour in the wet napkin. This colour is due to urate crystals which turn pink in air and this appearance is normal. The urine should not contain blood and should be virtually odourless apart from a slight ammoniacal smell. On the other hand if the urine has a fishy or stale smell with or without frequency check for a urinary tract infection

The stream is normally forceful but a weak dribbling stream in a male infant is suggestive of obstruction to outflow because of urinary valves and needs prompt assessment.

Vomiting

Babies spill or posset and this is so common as to be normal especially if there are no other associated symptoms. Spilling, coughing, gagging or ruminating during or after a feed is due to a combination of overt and/or silent gastro-oesophageal reflux. If there is also discomfort causing writhing movements of the neck and body then heartburn (oesophagitis) caused by inflammation of the oesophageal mucosa by refluxing stomach contents is the reason. Sometimes blood may be brought up with vomiting this being caused by small mucosal erosions in the oesophagus. These symptoms are exacerbated if baby lies in the supine position and they can be relieved by placing baby on the right side or else prone.

Projectile vomiting with discomfort during or immediately after a feed and becoming more frequent as the days pass suggests pyloric stenosis. Baby is usually male and weight gains are minimal or nil. Pyloric stenosis is insidious and presents in the first few weeks (usually 4-6 weeks).

Vomiting may be part of gastro-enteritis in which case the bowel motions should be loose, mucousy, explosive and frequent. If the motions contain flecks of blood consider a bacterial gastro-enteritis. On the other hand no blood is seen with a viral gastro-enteritis.

Bile stained vomiting and a distended abdomen means a bowel obstruction, while effortless clear vomiting on the first day or two and with no abdominal distension is likely to be due to duodenal atresia. Both of these conditions require urgent surgical attention.

If vomiting is associated with lethargy, irritability or fever, consider any form of septic illness

Weakness of a Limb

If an arm or leg seems to have become weak with limitation of movement or is obviously flaccid then a bone infection (osteomyelitis), joint infection (septic arthritis) or a fracture should be considered and urgent medical attention would be needed. With a weak or flaccid limb with or without a fracture always consider child abuse (battered baby).

Weight gains

Expect baby to regain birth weight at about 7-10 days. After that baby will gain 180-210 grams (6-7ozs/week) - an ounce a day except Sundays. The birth weight is doubled by 5 months and tripled by 15 months. These figures only apply to babies of average birth weight.

Chapter 19 : Conditions for discussion or investigation

Vesico-ureteric reflux

The association between vesico-ureteric reflux and kidney damage was recognized over 40 years ago. Since then much work has been carried out in Christchurch in investigating this association.

Vesico-ureteric reflux occurs when urine is refluxed from the bladder up one or both ureters towards the kidneys. At present it is graded into 5 categories

- grades 1 and 2 when a whiff of urine from the bladder is squirted into the ureters only.
- grade 3 when urine is refluxed up to the area of the renal pelvis but does not enter the kidneys.
- grade 4 & 5 when urine is refluxed into the kidney substance. Grade 4 & 5 reflux have the potential to cause kidney damage especially if infected urine is refluxed.

It is a familial condition and consequently if an older sibling or parent has been affected then the new baby should be investigated. These investigations include an ultrasound study of the kidneys and a micturating cysto-urethrogram (MCU) at about 6-8 weeks of age. The ultrasound will determine the size and shape of the kidneys and may be able to detect obvious scarring. The MCU will determine if urine is being refluxed or not and if so what grade of severity. Any damage done to the kidneys as a result of significant vesico-ureteric reflux occurs in the first few years of life.

All grades of reflux will resolve spontaneously with time but if significant reflux (grades 4 & 5) is present baby should be placed on prophylactic anti-microbial medication for up to one year and the urine checked at regular intervals or if baby at any time appears unwell. The aim is if possible to prevent kidney damage and there has been some debate over the respective advantages of either long term anti-microbial medication while waiting for natural resolution to occur or surgery with re-implantation of the refluxing ureter into the bladder wall to prevent further reflux from occurring.

Foetal alcohol syndrome

The foetal alcohol syndrome (FAS) was first described over thirty years ago in a small group of unrelated babies who had been born to mothers with chronic alcoholism. These babies were small and had small heads (microcephaly). Later on they became developmentally delayed and showed fine motor co-ordination problems.

At birth the physical findings may not be very obvious but they include shortening of the elliptical space between the eyelids (palpebral fissures), mid face flattening (hypoplasia). a shallow vertical groove in the upper lip (philtrum), a thin upper lip, small size and a small head circumference   These clinical features gradually lessen with time but the small size and head circumference along with varying degrees of developmental delay persist. FAS is a major cause of mental retardation. Arithmetical deficits are characteristic along with poor judgment, distractability and difficulty in the perception of social cues. Hearing disorders with recurrent glue ear or sensori-neural loss are fairly common in these children.

The full FAS is seen only in infants born to mothers with a heavy daily intake of alcohol. The syndrome is not thought to be related to associated malnutrition but rather to alcohol or its breakdown product, acetaldehyde. There is divided opinion as to how much alcohol and when is safe for the pregnant mother. It should be remembered that there is rapid development and brain growth in the first 8-12 weeks of the pregnancy as well as a further brain growth spurt in the last 2 months and therefore these two periods are vulnerable times for the foetus.

For there to be a major risk to the foetus mother has to have a daily alcohol consumption of 6 or more drinks per day, or at least 5 drinks on a single occasion and at least 45 drinks a month. It is accepted by most that somewhere between 1 drink per day to 2 drinks per week during the first 2-3 months of the pregnancy is not likely to be associated with foetal malformations. However mothers should not become intoxicated and if they have 2 drinks in a day then they should be spread over 2 hours. Total abstinence is not necessary.

Down syndrome

Described by John William Down in 1866

Down syndrome is seen in 1 in 850 live births and in the pregnancies of 1 in 340 mothers over 35 years of age. Down syndrome is usually (but not always) recognizable at birth. Look particularly for an upward slant to the eyes, flat head (brachycephaly), small ears, thickened neck folds, small slightly down turned mouth, relatively large tongue, Brushfield's spots in the iris, simian palmar creases, a large space between the first and second toes and general muscle hypotonia.

There can be several associated complications with Down syndrome. About 6% will have congenital hypothyroidism and subsequently periodic screening should be undertaken to exclude later thyroid dysfunction. Much has been made of neck problems with atlanto-axial instability of the cervical spine and resulting compression of the spinal cord although about 85% of children with Down syndrome show no evidence of atlanto-axial instability. It is more likely in males older than ten years but one should be wary of hyper-extending the neck of any child with Down syndrome. A cervical spine X-ray can be taken at 5-6 years of age but this really has no bearing as to whether the child can take part in any physical sport. There is in fact more risk of neck hyper-extension during intubation for a general anaesthetic. The upper cervical spine width is also shorter in Down syndrome than in the normal population.

Growth velocity begins to slow down between 6 months and 3 years of age and this is also the time when intelligence falls. At 1 year the average IQ is about 70 but declines to about 40-50 by 3 years. During this time of deceleration in growth and intelligence, head circumference also slows down. However with enrolment in an early intervention program significantly higher scores can be obtained in intellectual and adaptive functioning and these children do not show the same rate of decline.

The older person with Down syndrome has a propensity for developing a neurological condition resembling Alzheimer's disease. This dementia usually occurs after 40 years and is seen in about 30% of the older Down syndrome patients. Their usual cheerfulness and affection is replaced by solemnity and a decline in personal habits and cleanliness. This form of dementia seems to be concentrated in families who have been most severely affected by early onset Alzheimer's disease. Alzheimer's disease is related to aging in the normal population and the dementia of Down syndrome seems to be related to the precocious aging of Down syndrome. There is a higher incidence of mothers with graying hair and an increased risk where mothers were under 35 years at the time of the birth of their baby with Down syndrome. There is no increased risk in mothers who were over 35 years or fathers who have had a Down syndrome baby when compared with the general population.

Congenital heart disease is common in Down syndrome and all new born babies regardless of whether a heart murmur has been heard or not should have an ultrasound study of the heart or an echocardiogram within the first week or two after birth. In the absence of significant heart disease the average age of death in Down syndrome has risen from about 25 years in 1983 to 49 years in 1997.

With more sophisticated screening tests in the first trimester where four markers are used, the detection rate for Down syndrome is about 90% and with the false positive rate reduced to 2-3%. The need for first trimester chorionic villous sampling carries a slightly higher risk of pregnancy loss than does a second trimester amniocentesis but either of these two tests will give a definitive answer

Thyroid metabolism

Thyroxine (T4) primarily affects foetal nerve cell differentiation, as well as the formation of neuronal processes and synapses and most of these events are occurring by the beginning of the second trimester.
Maternal thyroxine does not cross the placenta in sufficient amounts to protect the foetus. Functional foetal thyroid tissue begins to appear between the 9th and 12th weeks of intra-uterine life. Thyroxine is needed for at least 6 months of foetal life and for up to 2 years after birth for normal nervous system and brain development. Foetal thyroid failure by the 5th month is indicated by radiological evidence of dental enamel hypoplasia and if there are absent epiphyses of the knee and ankle then thyroid failure has occurred by the 7th foetal month.

Congenital hypothyroidism

Screening tests for congenital hypothyroidism can be confusing and the results will depend on which day in the baby's first week the Guthrie test (a screening test for a number of rare conditions including thyroid dysfunction) is performed. Using a combination of thyroxine (T4) and thyroid stimulating hormone (TSH) values in newborn screening, congenital hypothyroidism will be found in about 1 in 6000 births.

If the T4 value is low and TSH value elevated (over 40mcU/ml), then the infant will have congenital hypothyroidism in about half these instances. If the T4 value is low and the TSH value is low normal then consider secondary hypothyroidism due to failure of the hypothalamic – pituitary axis.

If the T4 value is low or low normal and TSH is normal then a thyroxine binding globulin deficiency which is quite benign is probable.

False positive results can be the result of too early collection of the Guthrie test blood sample. TSH values peak at 30 minutes and then decline over the first 24 hours finally reaching normal values over the next 3-5 days. T4 levels are relatively low at birth and slowly increase to a peak over 24-36 hours and then slowly decline to normal values over a period of weeks or months. Remember also that transient hypothyroxinaemia of prematurity will show low T4 levels and slightly raised (but below 20mcU/ml) levels of TSH. These values rapidly become normal over the next few weeks. If there has been intra-uterine exposure to antithyroid drugs including iodine, maternal antithyroid antibodies or there has been an endemic iodine deficiency then transient hypothyroidism in the baby may result.

There is minimal transfer of thyroid hormones from mother to foetus in late gestation because the placenta and foetal membranes rapidly convert maternal thyroid hormones into inactive metabolites. Mother's breast milk will contain T4 and T3 but in amounts insufficient to allow normal neurological development of the baby. The signs and symptoms of congenital hypothyroidism are usually minimal in affected babies at the age of 10-21 days and in most cases treatment has been commenced by 2 weeks. Clinically the most sensitive indicator of hypothyroidism in the newborn is the circulation and if the hands and feet are warm and well perfused then hypothyroidism is unlikely. Other signs and symptoms may be subtle but look for diminished activity, poor feeding, delay in passage of bowel motions, persistent low temperatures or prolonged jaundice. Clinically the tongue may appear thick, the features coarse and the posterior and anterior fontanels will be large due to delayed bone maturation.

Treatment is life long and should start as early as possible and consists initially in rapidly raising T4 levels into the upper half of the normal range and maintaining these levels for at least the first year.

Congenital hyperthyroidism

A sustained foetal heart rate of 160/minute raises the possibility of foetal hyperthyroidism. The normal foetal heart rate in the second trimester is 135-145/minute and in the third trimester it is 125-140/minute. Antithyroid treatment for mother carries the risk of neonatal hyperthyroidism with a goitre, but usually the newborn baby has normal thyroid function. Propylthiouracil will cross the placenta but is the preferred treatment for maternal hyperthyroidism in pregnancy because of its greater ability to bind with proteins and therefore result in less transfer across the placenta. Baby will not show any clinical abnormalities as a rule but there may be biochemical evidence of elevated TSH levels and T4 values will be in the low normal to normal range. In most cases the thyroid status of the baby is not related to mother's condition but to the treatment she has been given.

Mothers taking propylthiouracil can breast feed baby without affecting the infant. The drug is not concentrated in breast milk and in the recommended dose for mother has no effect on the baby.

From a clinical point of view in the baby look for a rapid heart rate (tachycardia), a goitre and prominent eyes (exophthalmos). There may be frontal bossing and early fusion of skull sutures (craniosynostosis) also. Tachycardia can lead to heart failure.

Haemophilia

This bleeding disorder affects about 1 in 5000 males. It is an X-linked recessive condition seen in males only and is inherited from an asymptomatic female carrier. With haemophilia there is a deficiency in factor VIII which is a factor in the normal clotting cascade. Factor VIII circulates in the plasma as a complex of two separate peptides – factor VIII coagulant protein, the function of which is to generate fibrin and von Willebrand factor (vWF) which transports factor VIII and in addition facilitates the adhesion of platelets to vascular endothelium. In haemophilia factor VIII levels are low but vWF levels are normal.

In the newborn severe bleeding due to haemophilia is unusual and when seen commonly presents as a large cephalhaematoma, gastro-intestinal haemorrhage, bleeding from the umbilical stump or bleeding after circumcision. A normal delivery is thought to be safe but if there is any anticipated difficulty or assistance such as forceps or vacuum extraction then delivery by Caesarean section should be undertaken.

The normal level of factor VIII is 100% (50%-150% range). Severe haemophilia is less than 1% factor VIII and is characterized by spontaneous haemorrhages. Values between 1%-5% indicate moderate haemophilia and usually mean bleeding from trauma while levels of 5%-25% are mild and bleeding only occurs with severe trauma.

Von Willebrand's disease

This is a family of bleeding disorders affecting both males and females and is caused by some abnormality of von Willebrand factor (vWF). It is an autosomally dominant condition being passed down generations and is due to decreased levels of vWF. This results in lower levels of factor VIII coagulant activity and diminished platelet adhesion to areas of endothelial injury. It is difficult to diagnose von Willebrand's disease in the newborn period because since it is an acute phase reactant the stress of the delivery may cause a transient rise in vWF. Definitive testing therefore should be delayed for at least a month.

Thrombocytopenia

Neonatal alloimmune thrombocytopenia

In this condition the platelet count in baby is low. The clinical spectrum is variable ranging from an increasing and disseminated petechial rash, mucosal bleeding to intra-cerebral bleeding and with any of these presentations an associated low platelet count.

In this particular condition the foetal platelets carry an antigen which has been inherited from the father and which is lacking in the maternal platelets. During the pregnancy foetal platelets enter the maternal circulation and stimulate the production of maternal antibodies against the foetal platelet antigen. These antibodies are then transferred across the placenta and destroy the foetal platelets. Foetal-maternal platelet incompatibility is present in about 1 in 50 pregnancies but the incidence of thrombocytopenia occurs in only about 1 in 2000-3000 babies.

About 50% of cases occur in the first pregnancy and once this has happened there is a 70-80% probability of recurrence in subsequent pregnancies. These subsequent babies should be delivered by Caesarean section. The platelet count in the infant may remain low for 4-8 weeks since it takes that time for the maternally derived antibody to clear. If baby is treated with serologically compatible platelets the platelets survive for 7-10 days while incompatible platelets would be rapidly destroyed.

Meningitis

- Is baby lethargic or irritable?
- Check the anterior fontanelle with the baby upright and sucking on a finger or a nipple. It should not be tense or bulging. If it is depressed or soft and pulsating then meningitis is unlikely.
- Is baby vomiting especially if this has not been a feature in the past?
- The neck may or may not be stiff.
- If there is a scattered haemorrhagic skin rash (petechiae) or more floridly blot like skin haemorrhages then sepsis is probable.

Chapter 20 : Possible early infections for baby

Group B Streptococcal infection

Neonatal group B streptococcal infection takes two forms; the acute septicaemic form occurring before 10 days of age and usually associated with maternal obstetric carriage of Group B Streptococci with or without complications and the delayed onset form coming after 10 days and not associated with maternal complications.

Group BIa streptococcal subtype is associated with the septicaemic early onset disease and manifested by respiratory distress and apnoeic spells in the new baby and with a rapid downhill course while the late onset illness is usually meningitis and is associated with subtype Blll. About 15% of women carry group B streptococci Bla subtype vaginally or rectally while subtype Blll is not present in the vaginal canal. Males have an equally high rectal carriage rate. Thus early onset disease is acquired intrapartum and the late onset disease from sources other than mother. It is important to note that pulmonary invasion by Group B streptococci can produce hyaline membranes and thus the acute respiratory form will mimic respiratory distress syndrome.

The risk factors for Group B Streptococcal infection are young maternal age, vaginal colonization, low birth weight (<2500gms), prematurity, a previous baby with group B streptococcal sepsis and prolonged rupture of the membranes (>20hours). Therefore the colonized pregnant woman in premature labour or with prolonged

Chapter 20 : Possible early infections for baby 157

rupture of the membranes is at risk. Also the concentration of protective maternal antibodies increases with increasing maternal age and this could explain why young maternal age is a risk factor for infection.

The prevalence of group B vaginal streptococcal carriage at the time of delivery ranges from 5 - 30%. If mother is a carrier for group B streptococci about 50 -70% of newborn babies are colonized. About 1 - 8% of these colonized infants develop severe sepsis.

Women who are group B streptococcal carriage positive and have at least one risk factor should receive intrapartum antibiotic prophylaxis. Risk factors include fever, amnionitis, premature labour, and if the membranes have been ruptured > 6hours.

The risk of infants developing early onset disease is lower if all women are screened rectally and vaginally at 37-38 weeks and intrapartum prophylaxis given to all positive women at 2 hours before delivery.

Staphylococcal skin lesions

If blisters are seen on the skin particularly about the wrists think of sucking blisters which will have occurred in the latter part of the pregnancy when baby has been sucking in utero. They may have a linear distribution.

But also and of far more significance blistering can be due to a superficial Staphylococcal infection where small or large flaccid blisters or blebs are seen. They rupture leaving a reddened base and treatment will require systemic antibiotics and strict skin asepsis

The blisters of Herpes simplex occur in clusters while those due to Staphyloccocal aureus as a rule do not.

Nail bed infection is not uncommon in the newborn baby. The tip of the finger will be slightly swollen and reddened and a pustule will be noticed at the base of the finger nail. Local asepsis and a course of oral Flucloxacillin should bring resolution.

Rarely there may be widespread bullae or blistering affecting most of the skin surface. The skin sheets off if sideways pressure is exerted (Nikolsky sign). This condition is known as toxic necrotising epidermolysis (Ritter's disease) and is due to a particular Staphylococcal aureus infection . Urgent intravenous antibiotics are indicated as well as strict asepsis for baby and staff.

Even more dramatic and with large flaccid bullae affecting the whole body is the very rare condition Epidermolysis bullosa letalis, which while it can be treated carries a very poor prognosis.

Rubella (German measles)

The frequency of congenital rubella after maternal infection with a rash varies depending on in which trimester the infection contracted. During the first 12 weeks of pregnancy it is about 80%, 50% at 13-14 weeks and 25% by the end of the second trimester.

Rubella defects will occur in all babies infected before the 11th week (usually congenital heart disease and deafness) and will affect 35% of those infected at 13-16 weeks (deafness alone). No defects will be seen in babies infected after 16 weeks. Intra-uterine growth retardation occurs in all babies infected early in the pregnancy and this growth retardation continues after birth. Babies infected during the third trimester may show growth retardation but this is not ongoing after birth.

Of all babies affected by the congenital rubella syndrome and who survive after the first year into adult life, all will have a hearing impairment which is usually severe, half will have eye defects usually related to microphthalmia or cataracts and they will show growth retardation of varying degrees.

The full blown congenital rubella syndrome consists of cataracts, congenital heart disease, microcephaly, deafness and growth retardation.

Before the introduction of rubella vaccines rubella epidemics occurred at 6-9 year intervals. Immunity conferred by natural infection even if occurring in childhood is sufficient to protect the foetus although it may not protect against maternal re-infection. The re-infection rate following the administration of rubella virus vaccines is higher than that after natural infection but nevertheless is not thought to pose a threat to the foetus or of spread to contacts.

Varicella (Chickenpox)

A distinction is drawn between congenital and neonatal varicella infection. Congenital varicella is rare and confined to those cases where varicella is transmitted to the foetus in the first or second trimester while neonatal varicella occurs when the infection begins on or before the tenth day of life which means that the incubation time for the disease occurred in utero.

Clinically the two forms of varicella are quite different in their appearance.

The risk of embryopathy (congenital varicella) following maternal chickenpox infection within the first 26 weeks of the pregnancy is about 2% with the highest risk lying between 13 and 20 weeks. Certain criteria must be met when establishing a relationship between maternal chickenpox infection and congenital anomalies in the baby.

(a) there must be evidence of maternal varicella infection during the pregnancy.

(b) congenital skin lesions corresponding to a dermatomal distribution are present on the baby.

(c) there must be immunological proof of varicella zoster infection based on either persistence of IgG antibodies in the baby beyond the normal maternal decline of seven months or the presence of IgM antibodies in baby after the delivery or the appearance of the typical chickenpox rash in a dermatome distribution several months after the delivery.

Fortunately congenital varicella is rare. The features at birth include hypoplasia of a single limb with cicatricial scarring in a dermatome distribution, microcephaly, cortical atrophy, cerebellatr hypoplasia, microphthalmia, cataracts and chorioretinitis. In addition necrotizing encephalomyelitis, dorsal radiculitis and pneumonitis may also be present. During the time of maternal viraemia the foetus becomes infected and it is likely that there is actually an episode of intra-uterine chickenpox.

Chapter 20 : Possible early infections for baby

During the foetal viraemia or secondary to viral multiplication in the skin the sensory ganglia become infected. The neurons in the sensory ganglia are susceptible to lytic infection with resulting cell destruction. This causes a loss in trophic function and leads to limb hypoplasia which is the most obvious component of congenital varicella.

Neonatal varicella requires an incubation period of 10-21days. In those cases where the maternal illness began 5 or more days before the delivery neonatal disease will appear within the first 4 days of life and the outlook for baby is good. Presumably this is because some of the maternal immunity developing has appeared before the delivery and been transferred to the foetus before birth.

In those cases where neonatal disease begins between 5-10 days after the delivery, then maternal illness has occurred from four days to one day before delivery and there is therefore little chance of any maternal immunity being transferred to baby. The placenta acts as a partial barrier to the transmission of the infection during the pregnancy and only a small proportion maternal infections result in neonatal varicella.

For practical purposes infants of mothers who develop chickenpox from 7 days before the delivery to 7 days after the delivery should receive zoster immune globulin (ZIG). If the baby develops severe disease then acyclovir (acycloguanosine) is given in a dose of 10-30mgm/kg/day for 7 days to the baby.

ZIG is prepared from patients recovering from herpes zoster, these patients being a source of high titre varicella zoster antibodies. (Herpes zoster and varicella are caused by the same virus). Recipients of ZIG have no detectable antibodies in their serum one month after immunization.

Passive immunization with ZIG will result in the prevention rather than modification of chickenpox. Because cell mediated immunity is increasingly suppressed during pregnancy there are good reasons for giving ZIG to any sero-negative pregnant mother who has been exposed to chickenpox.

Immunological evidence of foetal intra-uterine varicella infection is shown by IgM antibodies to varicella zoster in the neonatal period and later persistently high varicella zoster IgG titres at 1-2 years of age. Herpes zoster infection during the pregnancy is not associated with serious maternal morbidity or with intra-uterine foetal varicella infection as far as is known.

Neonatal Herpes

Herpes simplex virus HSV-1 (cold sores) and genital herpes virus HSV-2 are cross reactive in all the standard serological tests. Most people have had HSV-1. There is a need to differentiate a primary (first time) infection from a non-primary (recurrence) infection since morbidity and mortality are associated with primary infection both with HSV-1 or HSV-2. Most (70%) of neonatal infections are produced by HSV-2 and the remainder by HSV-1. About 90% of these infections occur after exposure of the infant to the virus at the time of vaginal delivery while intra-uterine infection is uncommon.

About half the infants exposed to a primary maternal HSV infection at delivery will contract neonatal herpes. By contrast less than 5% of infants exposed to a re-activated maternal infection will be infected.

If a mother is experiencing her first attack of genital herpes at the time of delivery or has an active exacerbation of recurrent genital herpes then it would be sensible to deliver by Caesarean section. Infants delivered vaginally from mothers who have had recurrent genital herpes are at low risk if mother is asymptomatic at the time. In recurrent genital herpes the presence of high titres of neutralizing antibodies to herpes contributes to the low attack rate in the newborn baby.

Neonatal herpes may involve the skin, eyes and mucous membranes in about 40%, the central nervous system in 33% or be widely disseminated in the remainder of cases. The onset of the disease is usually in the first or second week of life but can occur at any time from day one to the third or fourth week. The typical skin lesions consist of clusters of vesicles or pustules on an erythematous base and are usually located on a presenting part such as the head or buttocks or any area of skin trauma. Sometimes the lesions appear as large bullae and can then be confused with staphylococcal vesiculobullous infection. Central nervous system disease may not have any associated skin lesions.

Treatment consists of acyclovir 10mgm/kg 8 hourly for ten days and treated infants with skin, eye or mucosal lesions will shown no identifiable abnormalities a year later. However in the disseminated or central nervous system groups despite treatment the prognosis in terms of mortality or morbidity is bleak.

Chapter 21 : Behaviour

Biological instincts

Behaviour in the infant, toddler or child is governed by certain fundamental social and biological instincts. For babies and toddlers there are two instinctive biological requirements – hunger and thirst. As we get older a third biological instinct appears – sex. But for our purposes hunger and thirst govern the biological needs of babies and children. These needs will be met either appropriately or inappropriately. There is no child who cannot or will not eat or drink providing he is hungry or thirsty enough. Indeed these instincts are so overwhelmingly strong that when in a plane crash in South America many years ago the survivors were so hungry that they were compelled to eat the remains of the dead passengers. If lost at sea, cannibalism by the survivors is not unknown. Again under circumstances of extreme deprivation and in an attempt to relieve thirst, urine or seawater is drunk. These extreme measures in an effort to combat overwhelming hunger or thirst would in a completely different environment where food and water are available be seen as inappropriate. However appropriately or inappropriately these fundamental instinctive requirements have to be met.

These biological needs do not need to influence a child's behaviour unless some form of coercion is exerted especially when that child refuses to eat.

Social instincts

The three social instincts (companionship, self esteem and curiosity) which govern children's behaviour are seen in every human being irrespective of age, sex or status.

Firstly the need for company and this is met in children primarily by the parents and secondarily by siblings, grandparents, relatives and friends, but parents are the essential companions for babies, toddlers and children. Parents if you like are a form of social food and a separation from the infant such as being put to bed or even moving to another room will elicit crying in an attempt to maintain the contact – much like a hungry child being denied food. In the older age group separation anxiety is seen when it is time for bed or being left at play centre, kindergarten or even going to school. The child will engage in any number of ploys in an effort to prolong the time before the inevitable separation occurs. eg; crying, dawdling, feeling sick or refusal. Infants and children are more comfortable when they know where their parents are and especially if they are nearby.

The second of the social instincts is the need for self esteem or if you like prestige or power. The infant soon realizes that he or she is extremely important in the parents' eyes and that they are the focus of much attention and love. This is healthy and normal and satisfying but as the child becomes older learned behavioural and inappropriate responses occur such as tantrums, aggression or destruction when a child is frustrated or thwarted in any way. At these times the foolish parent will give in to the child's demands thereby enhancing in the child's mind a sense of his power and importance but of course in an entirely inappropriate way. He has now gained control over his parents. This of course is very satisfying to his sense of self esteem and will then spill over into other areas such as refusing requests to eat or go to sleep or to use the toilet.

These latter three areas (eating, sleeping and toileting) are domains in which the parent is incapable of forcing the issue and will invariably lose the struggle and again enhance in an inappropriate way the child's sense of power.

The third social instinct is that of curiosity and this is seen in the need to explore surroundings, to have different experiences and to be entertained. It is realized in the toddler/child age group by sensible involvement with their parents allowing the child to explore surroundings, develop safely and to acquire independence but at the same time protecting the child from dangerous although interesting activities such as power points, heaters jug and iron cords and high ledges etc.

Thus the fundamental instinctive needs for babies and children both biological (eating and drinking) and social (companionship, prestige and curiosity) are met primarily through the parents. As the child grows and develops into adolescence, the needs remain the same but the manner in which they are met come from other sources as well. But the foundation stones have been laid down by the parents.

The key to childhood behaviour is for parents to accept that these fundamental social and biological urges are present and real and to meet them sensibly, fairly and with absolute consistency. Certain limits therefore are set in terms of behaviour and the needs met appropriately rather than inappropriately. It would therefore be quite inappropriate to accept without attempting to modify anti-social behaviour such as temper tantrums, willful destruction and physical or verbal aggression. These reactions are all attempts by the child to exhibit the need for prestige or power and to exert control over the parent and reverse a decision that he or she has not liked.

Disciplines

There are three techniques that can be employed and while two of these will fail they are nevertheless attempted by many well meaning parents. The first technique is to smack or beat the child and which if employed frequently is inappropriate and hardly conveying to the child that aggression is the right behavioural response.

The second technique is that of extinction or of taking no notice of the behavioural problem. This technique is very useful and effective in dealing with attention seeking behaviours which while they may be irritating in themselves are not socially dangerous or inappropriately aggressive. But for tantrums, aggression and destruction this technique is unhelpful.

Chapter 21 : Behaviour

The third and final and indeed the only effective management for unacceptable behaviour (tantrums, aggression, destruction) is that of isolation or "time out". This is practiced in the first few years of life and if it is to be helpful certain principles must be applied with absolute consistency and applied only for the above behaviours.

The moment the unacceptable behaviour (tantrums, aggression, destruction) occurs, the parent in a neutral and calm voice but nevertheless acting immediately takes the child and places him/her in a neutral non interesting space such as a hallway, a spare room, toilet, or the laundry but NOT the bedroom. There is no reasoning with the child or a second chance offered. Calmly the parent says something like "I am putting you in this room because of your such and such behaviour. The child is not smacked or shaken, the voice is not raised and the door is not slammed. The child is then ignored with no form of contact for a specified period of time – one minute per year of age. Pleading, cajoling, screaming or kicking walls does not elicit any parental response.

At the end of the specified period of time the door is opened no matter what the child is doing and the parent says calmly "you may come out now".

The child will make one of about four responses, namely to continue with the outburst in the neutral space, to sulk in the neutral space, to return to the familiar surroundings and be angelic or biddable or return to the familiar room and re-commence the tantrum. Should the child continue with the behavioural outburst either in the ordinary room or the neutral "time out" room he/she is immediately put back with the same calm explanation and the door locked again. Should the child discontinue the behaviour (biddable, sulking) the matter is then ignored and on no account does the parent apologise for what has happened. If these limits are observed and the approach is absolutely (100%) consistent the child's behaviour will initially deteriorate in further attempts to exert control over the parent. However after this difficult period and once the child realizes that his parents are resolute and completely consistent without letting any of the unacceptable behaviour pass unchecked a gradual and steady improvement will be seen.

It can be seen that with this isolation technique the child's instinctive needs for company, importance and curiosity have been temporarily suspended because of the unacceptable behaviour. In other words the child learns that there will be no parental company, he has not been able to exert control over his parents, power being taken out of his hands and in a neutral space there are no interesting objects to take his curiosity. Remember should this technique of "time out" be used then it must be accompanied by "time in". Here if there has been at any time a period of "good" behaviour even if only briefly then this is recognised by the parent who says something like "you are behaving very well (not "you are good") and we are going to do some interesting thing together or go somewhere. This shows the child that additional company, praise and interests are being met for him/her by one of the parents and all because of acceptable behaviour rather than a period of isolation (with suspension of social needs) for unacceptable behaviour.

Once the child realizes that his parents are completely consistent and fair and of course each parent must back up the other at all times when it comes to management, the child will become biddable, secure and confident. In other words he is meeting his social instinctive requirements especially for importance and control appropriately rather than inappropriately. The child then learns that when his parents say something they mean it, and will not be changed by tantrums and so on. This will spill over into other areas of child management. However it is quite inappropriate to attempt "time out" techniques when managing eating refusal, sleeping patterns and toilet training. These are areas where the parent cannot coerce the child in any way and management will have to be different.

Sleeping

For the first few months and perhaps year or two of life, sleeping problems are managed differently – smacking, extinction and time out are foolish and a waste of time and will fail. As is known the infant has an overwhelming need to meet social instinct No. 1 for parental company and will see being put down for sleep as a traumatic separation.

Chapter 21 : Behaviour

Once asleep infants unlike adults wake frequently and may find the absence of a parent stressful and crying will ensue. There are two schools of thought one of which is to let the baby or infant cry and not go near the infant.

While after several days or weeks this may seem to be a cure it is very upsetting for mother whose natural instinct is to attend to her baby. If baby is ignored this will do little for baby's sense of security in terms of company for in the future, a parent may or may not be present when needed. A similar analogy is to refuse the infant food knowing that the child is crying and hungry.

The second and sensible and more labour intensive approach is to accept that baby is crying and for whatever reason. A parent (usually mother) must go into the room to see what the problem is - lonely, hungry or in pain? Given that the baby or infant must be attended to, some sort of practical programme should be followed and followed consistently night after night. This will be tiring but in the long term a secure child will emerge.

After a quiet time before bed baby is put into the cot. Mother will stay with baby with her hand on baby but no other interaction until baby falls to sleep - no apparent separation for baby. A soft light may be left on and/or the radio left playing very quietly especially if baby is used to noise during the daytime. Mother leaves the room. The next time baby wakes and has been crying for about one minute a parent must go into baby's room. She must be neutral and is prepared to feed baby, change napkins but not to play games or to leave the room with baby. Baby is then replaced in the cot and mother stays with baby with her hand on baby until baby falls asleep again. Then mother quietly leaves the room. These babies with separation anxiety usually wake several times each night. In each event the same procedure is followed exactly except that for successive waking and crying an extra minute is added working up to about four to five minutes but no longer. Too long a period of absence will make it very difficult to settle a distraught baby even with a parent present.

As can be seen this is tiring for parents and in some cases parents will become cross or angry with baby and may become aggressive towards baby. If this stage is reached and before any damage is done to baby, then the infant is given an appropriate sedative (one of the antihistamines) each night for seven consecutive nights. This allows both baby and parents to have an uninterrupted sleep and to recover. Management then continues in the same cyclical way until the desired result follows. This approach while tiring shows the infant or baby that a parent is always present if needed and when separation becomes too much and that the instinctive need for company has been met. This will later engender security and the ability to tolerate future separations more easily.

Eating

Smacking, extinction or "time out" are again a waste of time and quite foolish.

For the toddler or older child it is usually the evening meal that is the problem when attention seeking inappropriate behaviour (importance) with more people present is more likely to achieve results.

It is a biological instinct to eat if hungry. The child if old enough is given the same meal as others present at the table and is not asked what he/she would like to have - it is a parental decision, not the child's decision. A small helping is placed on the plate and no cajoling or threats displayed. The child is not to learn that his or her eating habits are worrying for the parents, since this will bolster inappropriately his or her sense of self esteem.

Once the first course is finished by those present, the child's plate is removed without any comments. If there is to be a second course (pudding,dessert), a small helping is given to the child even if the first course was not finished.. A punitive parent will say there is no pudding because you did not eat the first course, and an indulgent foolish parent will say here is a big helping of pudding since you must be still be hungry, because you did not have much before.

If this attention seeking child then has a tantrum or displays any other angry or aggressive behaviour because they did not have what they wanted at the meal then the isolation ("time out") technique is immediately put into action.

Thus unacceptable behaviour in an effort to meet their instinctive social needs can be avoided.

Anxiety over separation (no company – social instinct No.1) when put down for a sleep, mother leaving the room, answering the telephone, being left at play centre or kindergarten or going to school can be seen as understandable.

Behaviour such as tantrums, aggression or willful destruction, can be seen as an effort to meet social instinct No.2 for importance (self esteem, power)

The child's innate sense of curiosity (social instinct No.3) should be encouraged but at the same time taking care that the child is protected from dangerous situations.

The key to successful and rewarding management of behaviour is to understand that every one has these social instincts and that in babies and toddlers they are met through parents. Their instinctive needs must be met sensibly and if they are not, be assured that they will be met and met inappropriately, now and later in adolescence and adult life.

We all, no matter who we are, what we do or what age or sex have needs for company, a sense of importance and a sense of curiosity. It is not too much to say that these needs influence in one way or another all of our activities.

Chapter 22 : Practical advances in the care of babies

Potable water and breast feeding

Probably the greatest influence on human health universally has been the acquisition dating from Roman times of potable water. In the early 1900's in New Zealand the infant mortality rate was very high – about 3 infants died each day mostly from "summer diarrhea" (infant scouring). By promoting breast feeding for the first 9 months of life and the "humanising" of cow's milk by using clean water adjusting the proportions of protein and carbohydrate to resemble breast milk, Truby King brought about a dramatic decline in the infant mortality rate. The ongoing emphasis on breast feeding ("breast is best") has continued and to the benefit of baby's health and intelligence.

Vaccination

Smallpox was a scourge and killer in the 18th century. The first attempts at vaccinating against this disease consisted of pricking the serum from the sores of smallpox subjects into the skin of another person to produce resistance resulting from hopefully a mild case of the resulting illness. The father of all subsequent vaccination programs, Edward Jenner in 1798 treated smallpox by inoculating subjects with fluid from the sores of vaccinia a disease of cattle, this vaccination having none of the dangers of the earlier attempts with the inoculation of smallpox material itself. This conferred immunity on the inoculated subject.

Louis Pasteur (1822-95) discovered the "germ" theory to explain diseases and from there later went on also to establish the fundamental principle that the injection of attenuated cultures of an organism would afford protection against the disease caused by that organism. Thus the works of Jenner and Pasteur were the earliest examples of the active production of immunity. The discovery of microbes and the recognition of antibodies in the blood of a person sick with an infection, the finding of a specific antibody directed against that "germ" in the blood allowed the determination of which organism was the cause. From there with increasing rapidity came the discovery of viruses, the development of antibiotics and the increasing refinement of immunology. Thus vaccination, the understanding of natural and acquired immunity and antibiotics have proved to be the cornerstones of refined medical management in babies and infants. These days the failure to use antibiotics for known conditions or the failure to have baby and child immunized would be an abrogation of one's duty as a parent.

Resuscitation techniques

The resuscitation of the asphyxiated baby has changed over the last two generations from the holding up of baby by the legs, to a few drops of Vandid on the tongue and intranasal oxygen. The "flat" limp blue-white asphyxiated baby is suctioned, intubated and given oxygen along with appropriate drugs.

In my time in Paediatrics which extended over 35 years and in the resuscitation of several thousand babies probably the most dramatic change in the "flat" baby's condition was brought about by the intravenous administration of Sodium bicarbonate. An apparently lifeless baby would almost immediately exhibit spontaneous respirations become pink and develop a normal heart rate. Bicarbonate would reverse the metabolic acidosis which resulted from asphyxia and this along with intubation and oxygen would usually produce an active well baby.

Neonatal technology has made huge strides forward and previously when a very small baby would almost certainly have failed to survive it is now not uncommon for the very smallest and earliest of gestation babies to survive and in many cases without later sequelae.

Jaundice

The treatment of jaundice has been simplified with benefit to baby and mother. Prior to the introduction of Rhogam (anti D hyperimmune gamma globulin) exchange transfusions of baby's blood for significant jaundice were common, time consuming and carried a certain morbidity. In my experience as a registrar as well as my colleagues I would have performed a hundred or more exchange transfusions and often one jaundiced baby would require several sequential exchanges over the course of an afternoon. Today an exchange transfusion is uncommon especially for jaundice due to Rhesus incompatibility.

The advent of light treatment with the development of the overhead phototherapy unit and later on fibre-optic blankets simplified the management of jaundice. Previously jaundiced was managed expectantly and as bilirubin levels steadily rose eventually an exchange transfusion with separation of baby from mother would be needed. The phototherapy unit negated the need to transfer babies from one hospital to another and this form of treatment which required baby's eyes to be bandaged was improved upon with the introduction of the fibre-optic blanket. Here baby could be treated at mother's bedside and the eyes also left uncovered.

When I look back over the last generation or so baby care has immensely improved especially with the encouragement of breast feeding, appropriate use of antibiotics, better and more effective resuscitation techniques, neo natal technology, the sensible management of jaundice and early enrolment in the immunization program.

Index

Numbers refer to chapter.

abdomen – scaphoid 2
acrocyanosis – dusky hands and feet 2, 4
alcohol – foetal alcohol syndrome, maternal ingestion 1, 15
alpha – lactalbumin 19
Alzheimer's disease 15
ambu bag 2
amnionic sac tear 4
amniotic fluid 1
amputation 2, 4
angel's kiss 2
anti D hyperimmune gamma globulin (Rhogam) 14
antibodies IgG1, IgG2, IgG3 14
anticonvulsants 7
anus 2
Apgar score 2
apnoea 19
appropriate for gestational age (AGA) 2
ASD – heart murmur, inheritance, septal occluder 10
aspirin 10
asthma 20
atlanto-axial instability 15
atopic dermatitis 5
atrial septum 10
atrium – right 10
baby measles (roseola infantum) 20
bacon – nitrosamines – brain tumour 1

Index

Barlow manoeuvre 11
Beckwith–Wiedeman Syndrome 8
behaviour – time in, time out 21
behaviour – tantrums, aggression, destruction 21
behaviour management – smacking, extinction, isolation 21
beta glucuronidase 14
beta lactoglobulin – genetic variants 19
biliary atresia 20
bilirubin – encephalopathy, beta glucuronidase 14
biliverdin 7, 14
blue baby 10
blue light 14
bony dysplasias 6
bovine serum albumin 19
bowel motions – colour, consistency 20
branchial cleft 2
breast feeding – drugs 17
breast feeding – exclusive, winding 17
breast milk - enhanced secretion 17
breast milk – jaundice, inheritance 14
breasts – physiology 17
bronchiolitis 20
Brushfield's spots 15
cafe au lait patch 2
caffeine – coffee, prematurity 1
calcaneo valgus 2
caput 4
carbamazepine – embryopathy 1
casein 19
cetaphil 5
chicken pox – treatment 16
choanal stenosis/atresia 2
chordee 2
cromoglycate 19
circumcision – pain 12
circumcision – urinary tract infection, penile cancer, AIDS, epididymitis 12
cisapride (prepulsid) 18
clavicle – fracture 4, 2

cleft palate inheritance 7
clinodactyly – fifth finger 2, 11
clitoris 2
club foot – treatment 11
coarctation of aorta – heart murmur, inheritance 10
coccygeal dimples 2
Coffin Siris syndrome 2, 11
coloboma – iris defect 2
colour recognition 9
congenital dislocation of hips – inheritance, treatment 11, 2
congenital rubella 6
contact dermatitis 5
Coombs test 14
corneal dryness 4
corticosteroids 5
cotinine 20
cows milk protein - allergy, sensitivity, intolerance 19
cryotherapy 5
cryptorchidism – treatment 12
cutis marmorata 4
cyanosis 2
cytotoxics – embryopathy 1
dehydration 20
dehydration fever 2
dental malocclusions 20
diabetes mellitus 20
diaphragmatic hernia 2
digits – accessory 11
disaccharidases 19
disaccharides 19
disposable napkins 5
Down syndrome 4, 6, 9, 11
Down syndrome – investigations 15
drug addiction – heroin, morphine, methadone, cocaine, marijuana 1
ductus arteriosus 10
ductus venosus 10
dummies 3
eating – attention seeking 21

ectropion iris 2
Ellis van Creveld syndrome 11
embryopathy 1
encephalocoele 4
entero – hepatic circulation 14
Epstein's pearls 2
Erb's paresis 2, 3
esotropia 9
exotropia 9
external meatus 12
eye blinking 9
eye colour 2, 3, 9
eye tracking 9
facial recognition 9
fibre optic therapy 14
finger - 5th – accessory 2
fingernails – hyperconvex 11
fluorinated steroid cream/ointment 5
fontanelle – anterior, posterior, third 6
foramen ovale 10
fore milk 17
foreskin 12
frenulum 7
friction blister/ulcer 4
gastroenteritis 20
gastro-oesophageal reflux – frequency, propping, milk thickening 18
german measles (rubella) 20
gestational age 3
Gilbert's syndrome 14
glaucoma 5
Goldenhar's syndrome 8
grandmother theory 14
granuloma gluteale infantum 5
green light 14
haematocolpos 12
haematoma - sub galeal 4
haemophilia - severity 4, 15
haemorrhagic disease of the newborn 2

head circumference 6
head line 11
hearing assessment 8
heart line 11
heart murmurs 2
heartburn 18
hectic spot 7, 20
helix – upper part of ear 2
heme 14
hemi facial microsomia 8
hepatitis - neonatal 20
herpes – treatment 16
hiccups – diaphragmatic contractions 1
hind milk 17
hip dysplasia 11
Hirschsprung's disease 20
holding preference 3
Horner's syndrome 4
hydrocolpos 2, 12
hydrocohaematometrocolpos 12
hydrolysed formulae 19
hydrometrocolpos 2, 12
hymenal tag 12
hyperthyroidism 6
hypopigmentation 5
hypoplastic nail 11
hypospadias repair 2, 12
hypothyroidism 6
hypotonia 2
inferior vena cava 10
instincts – biological; hunger; thirst 21
instincts – social; company; self esteem; curiosity 21
intersex anomaly, 2
intertrigo 5
iris depigmentation 4
iron 7
keratosis pilaris 5
Klumpke's palsy 2
labia majora 2

labia minora 2
lactobacilli 7
Laerdal mask 2
large for gestational age (LGA) 2
laser therapy 5
let down reflex 17
life line 11
lipobase 5
lithium – embryopathy 1
long QT syndrome 18
lymph nodes 2
macule 5
Marfan's syndrome 8
McCune Albright syndrome 5
meconium 21, 2
meningitis 20
metoclopramide (Maxolon) 18
metopic suture 6
microcephaly 6
milia – sebaceous gland hyperplasia 2
miliaria – sweat retention cysts 2
minocycline 7
Moro (startle) reflex 2.3, 3
mouth asymmetry 4
nasal deviation 2
naso-lacrimal duct 2
neurofibromatosis 5
nipples – smell 3
Nissen fundoplication 18
nuchal fold fluid collection 4
nummular eczema 5
oculo-vertical reflex 2
oedema – fluid accumulation 2
omeprazole (Losec) 18
orbicularis oris hypoplasia 2
Ortolani manoeuvre 11
osteogenesis imperfecta 7
osteomyelitis 20
otitis media 20

Index

ovalbumin 19
ovo-testes 12
oxytocin 17
palmar creases 2
Pavlik harness 11
PDA - machinery heart murmur; closure 10
penile torsion 12
perianal erythema 5
perineal membrane 2
petechiae – pin point haemorrhages 2
phenylalanine 6
phenytoin – embryopathy 1
phototherapy 14
pilo-sebaceous follicles 4
placement (stepping) reflex 2
placenta – heat exchange 2
plagiocephaly 2
port wine stain 2
pre-auricular skin tags and dimples 2
preputial cyst – foreskin 2
processus vaginalis 13
prolactin 17
prop feeding 17
propylthiouracil – goitre 1, 15
pseudoainhum 4
pyloric stenosis – inheritance 18
quickening – foetal movements 1
ranitidine (Zantac) 18
recti abdomini divarication 2
red cell enzyme deficiency 14
red reflex 2
red wine 7
respiratory distress syndrome (RDS) 16
response to sound 3
retinoic acid embryopathy 1
retrognathia – positional 2
rifampicin 7
rubella – vaccine 16
sacral cleft 13

sacral dimples 2
Sandifer's syndrome 18
sclera – white part of eye 2
separation anxiety 21
sepsis 20
septic arthritis 20
septum primum 10
septum secundum 10
simethicone 19
simian crease 2.3, 11
skin desquamation 4
skin to skin contact 3
slapped cheek syndrome 20
sleeping – sedatives 21
sleeping times 20
small for gestational age (SGA) 2
smoking – small baby 1
social workers bruise 5
sorbitol 7
spermatogonia 12
spinal dysraphism 13
splenic enlargement 2
sternomastoid tumour 6
steroids – virilisation 1
stork bite 2
stratum corneum 5.13
streptomycin – hearing deficit 1
Sturge Weber syndrome 5
sub mucous cleft 2
sucking reflex 2
sucrose solution 3, 19
suture fusion/suture overriding 2
sweat retention cysts 4
Sydney line 2, 11
syndactyly – soft tissue 1
synostosis 6
talipes – structural/positional 2
taste preference 3
tears 9

teething – appearance 7, 20
telangiectases 5
temperament 2, 3
teratogen – embryopathy; foetal malformation 1
testes – retractile; cancer; torsion 2, 10
tethered cord 5
tetracyclines – discolouration of teeth 1, 7
thalidomide – embryopathy 1
thigh creases 2
thrush 5
thumb – bifid; sucking 2, 3, 11
thyroid dysfunction 15
thyroid stimulating hormone 15
thyroxine 15
tibial torsion 11
tide mark dermatitis 5
tongue tie 2
torticollis 2, 6
Treacher Collins syndrome 8
tridione – embryopathy 1
tummy time 6
tunica vaginalis 13, 13
Turner syndrome 10
umbilical cord 1
umbilical hernia 2
umbilical stump 2
Unna's naevus 5
urate crystals 2, 20
urethral valves 2, 20
urinary tract infection 20
uvula bifid 2, 7
valproate – embryopathy 1
ventricle – right 10
vernix caseosa 5
vesico ureteric reflux – inheritance; investigations 15
vestigial nipple 10
visual accuity 9
visual recognition 3
vitamin D deficiency 6

Index

vitamin K 2
voiding 2
vomiting – bile stained; clear 20
vomiting – hyperemesis gravidarum 1
vomiting – projectile 18
von Willebrand's disease 4
VSD – heart murmur; inheritance 10
warfarin 1
whooping cough 20
witch's milk 10, 20
xiphisternum 2

www.ingramcontent.com/pod-product-compliance
Lightning Source LLC
Chambersburg PA
CBHW060847170526
45158CB00001B/263